Supervisor's Foreword

As a Ph.D. student, Denitza Denkova joined our Institute for Nanoscale Physics and Chemistry at Department of Physics and Astronomy, KU Leuven, to investigate near-field optical properties of plasmonic nanostructures. In the framework of her Ph.D. research, she has been hard working and efficient in pursuing challenging problems in modern optics and, more specifically, in nanoplasmonics. In her Ph.D. work Denitza has discovered a series of new phenomena in several well-defined individual and coupled plasmonic nanoantennas. To be more specific I would like to comment on her remarkable work on visualizing magnetic hot spots in plasmonic nanoantennas.

Denitza, in collaboration with Niels Verellen, Alejandro V. Silhanek, Ventsislav K. Valev, Pol Van Dorpe, and me, has developed and realized the first direct visualization of the near-field magnetic field component (Hy) of plasmonic nanostructures by scanning near-field optical microscope (SNOM). This is clearly a major achievement in nanoscale imaging of both electric (by now a well-established technique) and also magnetic fields generated by nanoplasmonic structures. Needless to say, this finding will open new horizons in nanoscale imaging in modern nanoplasmonics.

The photonics community has long disregarded the magnetic component of the electromagnetic field of light, focusing mostly on exploiting the electric field. The reason is that at optical frequencies, natural materials do not exhibit strong magnetic response, so the magnetic field is, first, not of crucial importance for those materials, and, second, very difficult to measure. The situation changes dramatically for the new classes of artificial materials, such as optical nanoantennas, plasmonic devices, and metamaterials. They are typically designed to have a very strong magnetic response; therefore, accessing their magnetic field is of crucial importance for their exploration. While in free space, the magnetic field component can be extracted from the electric one, the confinement of the light in these novel materials prevents such a straightforward relation.

Quite remarkably, by imaging the magnetic field component, Denitza has resolved a very puzzling fundamental issue, which can be formulated in the language of a quantum mechanical "Schrödinger cat" problem: How many stripes does the "plasmonic cat" have? Surprisingly, looking at the stripes of "plasmonic cat" through an "electric field magnifying glass" (SNOM in the E- imaging mode) one can see three stripes, whereas when observing the same cat through a "magnetic field magnifying glass" one can see only two stripes. This surprising difference in the observed number of stripes turns out to reflect a profound natural difference in the boundary conditions for the electric and magnetic fields confined in the restricted geometries of plasmonic nanoresonators.

These new findings, forming the basis of Denitza's Ph.D. thesis, will have an important impact on our fundamental understanding of local magnetic and electric fields in plasmonic nanostructures, and also on the development of nanophotonic devices with promising applications in chemistry (catalysis, bio- and chemical-sensing), energy harvesting, and biomedicine (cancer treatment by hypothermia, early disease diagnostics). This work also offers a remarkable new way, different from conventional techniques, to investigate experimentally various nanostructures and also for studying materials showing strong magnetic interaction with light in the areas of metamaterials, chemistry, and all-optical chips.

I am sure that Denitza Denkova will continue to contribute very creatively to research in the fields of nanoplasmonics, photonics as well as optics and condensed matter physics in general. She is clearly a very talented young scientist who pursues scientific problems at a very high level of sophistication and I wish her all the best in her scientific carrier.

Belgium   Prof. Victor V. Moshchalkov
April 2015

Abstract

The interaction of light with nanoscale objects has been one of the hot topics in the field of photonics in the past few years. More specifically, one of the promising directions is the field of plasmonics, which deals with the interaction of light with metals. Illumination of metallic nanoparticles results in the excitation of so-called surface plasmons—collective oscillations of the free electrons in the metal, driven by the incoming light. At certain frequencies, different resonant modes can be excited, which results in significant enhancement and localization of the light in the close vicinity of the nanostructure—the particle is basically acting as an antenna at optical frequencies. Such local light field enhancements (also called *hot spots*) have shown promising applications in various areas, for example for single molecule detection, bio- and chemical-sensing, all-optical chips, cancer diagnostics and treatment, *etc*.

Further development of those applications requires detailed understanding of the full picture of the light–matter interactions. Since light is an electromagnetic wave, the response of a material to illumination is determined by the interaction of the incoming electric and magnetic fields with the medium. Thus, for the plasmonic structures, depending on the specific application, it is crucial to characterize the resonance wavelengths for the different resonant modes, the charge and current distributions in the particles, as well as the electromagnetic near-field distribution of the light in the vicinity of the structure. Different methods, each with its own scope, advantages and disadvantages, already exist for imaging most of those parameters.

Still, one of these parameters, namely the *magnetic* near-field distribution of photonic structures, has received far less attention from the optics society. The reason is that at optical frequencies natural materials interact mainly with the *electric* component of the electromagnetic field of light and negligibly weakly with the *magnetic* one. Therefore, the magnetic light–matter interactions are first, not of crucial importance for those materials and second, very difficult to measure.

However, recently new classes of artificial materials, so-called metamaterials, have been developed, which enhance and exploit these typically weak magnetic light–matter interactions to offer extraordinary optical properties. For example, such materials could refract light in a direction, opposite to the one in all conventional materials. This can allow realization of exotic devices, such as an invisibility cloak, flat lens with no resolution limit, *etc*. Therefore, in the last years, a need has appeared for both optical magnetic sources and for detectors of the magnetic field of light and tremendous efforts have been invested in this direction.

The main goal of this Ph.D. is to offer new experimental possibilities for near-field imaging of the magnetic field of light with sub-wavelength resolution and to implement the technique for studying different plasmonic nanostructures.

The experimental technique which we use is scanning near-field optical microscopy (SNOM). This is a method, based on the scanning of a probe in the near-field of the sample. Depending on the type of probe and measurement configuration, different field components can be accessed with sub-wavelength resolution. Imaging of the different *electric* field components is nowadays a well-developed procedure. However, it is still an ongoing challenge to experimentally access the *magnetic* field components, which interact much weakly with materials. Recently, it has been shown that the *normal* (relative to the sample surface) magnetic field component can be accessed by a split-ring aperture probe.

To fill in the last standing gap in the full electromagnetic mapping of the near-field of photonic devices, in our work we focus on imaging of the *lateral* (tangential) magnetic field component of the light. We demonstrate that the metal-coated SiO_2 hollow-pyramid circular aperture probe of a SNOM can be used as a detector for the lateral magnetic field of light. Moreover, we show that it can also be used as a tangential magnetic dipole source. We use the technique to study the lateral magnetic field of plasmonic structures with different geometries.

During the period in which our work has been carried out, the task for mapping the lateral magnetic field has been also addressed by other groups. They use a similar SNOM technique, which however utilizes other types of probes, based on optical fibers. The Si-based probes which we use have the advantage of being more robust and providing images with better topographical resolution, simultaneously with the optical image. A more detailed comparison is presented in the thesis.

In Chap. 1 we first discuss the interaction of light with metallic nanostructures and the resulting plasmonic effects in those structures. Several state-of-the-art applications are highlighted, which, naturally, require detailed characterization of the plasmonic structures. In the second section we explain the experimental difficulties for plasmonic studies and how the scanning near-field optical microscopy technique can overcome them. The final section discusses the physical origin behind the weakness of the magnetic light–matter interactions at optical frequencies. We also explain the specific SNOM setup which we use to probe those interactions and to image the magnetic field of light.

In Chap. 2 we experimentally demonstrate that the hollow-pyramid SNOM probe images the lateral magnetic field of light of different plasmon modes in plasmonic bars with different lengths. Supported by simulations, we describe how the coupling between the probe and the sample induces an effective magnetic dipole, which, during the scanning of the probe, efficiently excites the plasmons in the bar only at the positions of the lateral magnetic field maxima. Respectively, at those positions the absorption in the bar is enhanced and the transmitted light is reduced. This effectively results in imaging of the lateral magnetic field of the plasmonic structure, with reduced transmitted light intensity at the positions with high lateral magnetic field. By measuring different bar lengths at different wavelengths and identifying the respective plasmon resonant modes, we construct a dispersion relation (energy vs. wavenumber) graph. The dispersion relation shows the characteristic curving deviation from the straight line of the light in dielectric media. This confirms that the effects which we observe are indeed of plasmonic nature.

In Chap. 3, supported by simulations, we discuss the theoretical background behind the imaging of the magnetic field of light with the hollow-pyramid aperture SNOM probe. Additionally to the previous chapter, here we suggest that (i) the stand-alone probe might be considered as a tangential optical magnetic point dipole source without the need to invoke the coupling to the sample, and (ii) it might be considered as a tangential optical magnetic field detector. Hence, we suggest that the probe is effectively behaving similarly to a lateral magnetic dipole. This is demonstrated for a metallic sample, while the applicability of the approximation for studying dielectric samples remains to be verified. We experimentally demonstrate the equivalence of the reciprocal configurations when the probe is used as a source (illumination mode) and as a detector (collection mode). The simplification of the probe to a simple magnetic dipole significantly facilitates the numerical simulations and the understanding of the near-field images.

Until now, we have used basic plasmonic antennas—gold bars to validate the magnetic field imaging technique. In Chap. 4, we use the validated technique to obtain the magnetic field distributions of plasmonic antennas with other geometries. We study both simple plasmonic nanoresonators, such as bars, disks, rings and more complex antennas, consisting of assembled horizontal and vertical bars in different geometrical configurations. For the studied structures, the magnetic near-field distributions of the complex resonators have been found to be a superposition of the magnetic near-fields of the individual constituting elements. This effect was confirmed by both experiments and simulations. We propose the following explanation: even though a weak interaction between the building blocks is present, the simplicity of the resonance mode structure and the clear spectral separation of the different modes in the bars result in insignificant perturbations of the near-field profile at the resonance wavelength by the presence of the additional building blocks. These findings should facilitate the study of very complex antennas, which can now be described as a simple superposition of their elementary building blocks' near-field distributions

Acknowledgments

The Ph.D. journey in not a path one walks alone, neither the Ph.D. degree is a destination one reaches alone. In the next pages, I would like to thank all the people who helped me throughout my journey and made it possible for me to reach my destination.

I would like to begin by thanking my supervisor, Prof. V.V. Moshchalkov for giving me the opportunity to come to Leuven and work on the challenging, but interesting topic of magnetic near-field imaging of plasmonic structures. Thank you for the guidance on how the modern scientific world works, but also for allowing me to preserve my personal style of work and giving me the freedom to fit in this complicated world in my own way. Thank you for giving me the opportunity to play with this big toy—the SNOM. It was already in the beginning we discovered that the SNOM is a she-microscope—one with a difficult character and a lot of mood-swings. But in the end, after 4 years, I believe I finally managed to get some understanding with her.

Next, I want to thank my co-supervisors—Dr. N. Verellen and Dr. V.K. Valev. Niels, I believe me being your first Ph.D. student was an interesting and useful experience for both of us! I am very grateful for all your help with the sample fabrication, simulations, data interpretation, and all other scientific (and not only) activities. Thanks for having the patience to guide me and help me. Ventsi, I want to thank you for your help, especially in the difficult starting period of my Ph.D. Thank you for diversifying my work and broadening it beyond the SNOM by involving me in your second harmonic generation studies.

I would also like to thank one of the scientifically most important people for me—Prof. A.V. Silhanek. Alejandro, it was a great pleasure to work with a scientist like you! I really love your, at first sight simple, but provoking questions on all aspects of my work. Thanks for having the patience to discuss all the tiny and small details, on which I was losing my sleep. Thanks for being picky and provocative. Thanks for all the discussions—face to face, on the phone, while driving the car, via emails, thanks for always being here for me (especially on Fridays), ready to talk about everything! Thanks for supporting me in preserving my own working style and for helping me to keep a healthy balance regarding that. Thanks for inspiring me and boosting my enthusiasm when I needed that and thanks for keeping my feet to the ground when I was loosing direction. I would also always remember the philosophical discussions about life, which were for me sometimes as important as the work-related ones.

Special thanks also to all my Jury members—Dr. N. Papasimakis, Prof. K. Clays, Prof. P. Van Dorpe, Prof. A.V. Silhanek, Dr. J. Vanacken, Prof. V.V. Moshchalkov, Dr. N. Verellen, and Dr. V.K. Valev for the critical reading and the constructive comments on my thesis. It was a pleasure to feel like "a real scientist" having a discussion with all of you.

It was also very useful and pleasure for me to have scientific discussions with Prof. P. Scott Carney and Dr. Alexander Govyadinov. Interesting experience were the discussions with our colleagues form the engineering department of KU Leuven (ESAT)—Dr. Xuezhi Zheng, Dr. Vladimir Volskiy, and Prof. Guy Vandenbosch, with whom it was a real challenge to bridge the physics language and the engineering language.

Besides the people who were scientifically directly involved in my articles and the thesis, I would also like to thank the other people who had the patience and made the effort to critically read and/or discuss my work with me—Prof. Vesselin Strashilov, Prof. Joris Van de Vondel, Dr. Joffre Gutierrez, Dr. Ward Brullot, Dr. Pavletta Shestakova (that's my mom!), Prof. Nikolai D. Denkov (that's my dad!)…

Many thanks also to our always-in-a-helpful-mode secretaries, Monique and Liliane, and the guys from the workshop—all the three Fille's. I would also like to express my gratitude to Stijn Caes for putting me back on my feet in a few quite critical situations.

For good or bad, to me life means much more than work. So, now it comes the time to also thank all those people who extended my life in Leuven beyond work.

A big thanks to all the colleagues in the group who were trying to create a nice working atmosphere, sometimes even a smile on the corridor from you guys was enough to boost up my mood—Matias, Joffre, Lise, Joris, Misha, Paulo, Jun Li, Xianmei, Vladimir, Junyi, Dorin, Gufei, Johan! Thanks to my office-mates Paulo, Matias, and Jun Li—mainly for having the patience to re-direct everyone looking for me at the office to my lab-cave on the third floor. It is not because I don't like you guys that I came once per year to the office—just the SNOM was getting jealous every time I was not with her! Jian, keep it in mind, you have to be very careful with her!

I would keep a lot of warm memories also from the other colleagues from the department. Definitely from the "veterans", who taught me the rules of the game around Leuven—here I have no choice but to start with Mariela's name, and then continue with the other party-animals—Bas, Werner, Niels, Bart Raes, Cristian, Pieterjan, Sasha, Yogesh, Tom, Tomas, Jo (sorry Jo, you are already in the "veterans" category), Ataklti (you too, you have a baby already!!!), Joffre (you have always been here by default, I think). And also many thanks to "the younger guys" with whom we shared some great moments: Tobias and Claudia (you know that I am in love with the gezelligheid you create around yourselves, right?), Kelly and

Acknowledgments   xv

Tom (I still want to try the water skiing!), Barbara and Didier, Hanna and Miro, Matias, Mattias, Piero and Pía, Bart, Johanna, Hiwa, Manisha, Sérgio, Petar, Misha, Vera, Bas, Fille, Fille, Fille, Katrien, Bart Ydens, Zhe, Winny, Cédric, and others.

Since sports is a huge part of my life, I want to also thank all the people who had the patience to teach me new sports or to get trained by me or simply the ones who dared to do sports with me.

Historically, one of my first social activities in the group was to join the soccer team—those guys from the Physics Department had a special style of making friends with ladies…. No, seriously, thanks to Jo, Ataklti, Bas, and Joffre for daring to bring women to the soccer team! It was very relaxing to kick some balls with all of you guys—Petar, Daniel, Fabian, Piero, Matias, Bert, Tomas, Pieter, Troy, Mansour, and of course the engineers—Son (thanks for organizing!!), Ruben, Dries, Tijs, Damien, Carlos, Mats, Asim, Gonzalo, Paola, Vittorio, Giovanni, Tassos, Guilherme, Sam, and all other Messi's on the field.

The other sport which you guys definitely got me hooked into was the squash of course! I know you mainly invited me to join because you were too lazy to go by bikes and you needed a transport by my limo-mafia car, but still, we reached a good synergy! Thanks for having the patience to teach me to play, I am now really a big fan! I believe the main credits here have to go to Mariela (my first opponent ever and also the first one to ever let me win), Bas (most definitely my best squash performance ever was that long-rallies-show which unfortunately ended up 3:2 for you), the Fille's (pity that you guys got scared and stopped coming just when I was ready to beat you!), Jo (I would dare to challenge you now!), Ewald, Bart, Werner, Kelly, Tobias, Arnaud; to the new generation—Carlos (! Si senõr!, I will start getting those drop shots one day), Joffre (for bringing up the competitive mood on the field), Enric (I am sure I won two sets that day!!), Matias (sorry, but I have to say it—my first, and probably only one ever 11:0 win), and the brand-new generation—Mattias, Pierro, and Bart—I am relying on you guys to keep up the squash group playing! It was also a pleasure to play with the engineers—Vyacheslav (thanks for organizing!!), Laurens, Damien, Axel, Yas, Sofie, Lucie, Bertram, and the rest.

Now of course we come to the biggest passion—volleyball! Guys, I can really not believe how hooked and enthusiastic you became for the volleyball and especially for the beach volleyball! I am really, really proud of you—you learned the triangulation, the kitty, and so many professional volleyball tricks! Thanks to all of you who shared the moments on the "beach"—Matias and Hiwa (thanks to both of you for keeping the organization running), Joffre (being as lively as you are, you are doing great regarding most of the important elements—counting the points, reserving the fields, fetching the balls and the only thing you might still work on a bit is your reception), Carlos (you are sometimes too demanding!), Bas (you are loosing shape), Mattias (that pass is not that important), Johanna, Vera, Manisha, Sérgio, Jochen, Ivan, Nastya, Stas, Ivana, Jono, and others.

And Mariela, I can really not believe it that you convinced me to participate in the Bike and Run (I knew I should have gone home after the first Chouffke) and I can definitely not believe that I liked it! Of course this was partially because of the huge euphoria in the whole department and the fantastic T-shirts we had. But mostly, it is because of my personal coach and running-buddy—Mattias. Kiddo, I really had a great time, I cannot believe that we made it in 2:43:14, 12.5 km/h!!! And hey, Ironman… come on, really… a pot on the bike?! We had the best style of all the teams for sure! Maybe only Hanna's team was competing with that picnic basket.

There are also plenty of other memorable events, for which I want to thank all participants: the crazy summer trips to the Rotselaar lake; the skiing trip to Austria, the Christmas Markt in Cologne, the motorbike trips, that legendary Hawaiian party, the cozy board game evenings at Tobias and Claudia's fantastic apartment, the countless just-one-beer visits to the Oude Markt, the once-in-a-lifetime pantoffels night, the refreshing ice cream and coffee-breaks just (but not restricted to) when it was needed. And of course the blasting Ardennen weekend (I accept betting on how many times I use the word "Thanks" in the Acknowledgments)…!!!

Among those events is definitely the surprise-birthday party which you guys organized for me! It really meant a lot to me, especially at those emotionally very difficult times! Thank you!

I also want to thank here Ward and Maarten for sharing with me my best conference ever, and moreover the social events related to it! I had a lot of fun guys, I know I was not your dream-company (Maarten!!), but I hope you too managed to enjoy the trip! Oh, come on, in the end we didn't get lost and we didn't crash the car and I was not eating *that* slow and we saw alligators and it was great!!!

I would also like to specially thank Ward, Joffre, and Mattias for the priceless support during the last stage of the writing of the thesis! Ward, I really cannot imagine how I would have managed without your amazing organizing and motivating skills and Latex skills too. Thanks to Joffre and Mattias for keeping a piece of me alive during the writing by dragging me out for some sports and breaks and by bringing tones of (sometimes even healthy) food and cookies to the lab!

Talking about sports, many thanks to my volleyball teammates and coaches for the great volleyball trainings and games with the Velvoc team and with the Eurogirls team. Girls (and guys), it was really a pleasure to play with all of you—Assia, Jana, Tereza, Kristina, Anja, Claudia, Julia, Clelia, Iva, Vale. Kristina, Herki, Rodri, and all the rest—it was great playing with you too at the beach volley fields and at the tournaments!

Here I would also like to thank all the other "local" people with whom we enjoyed different activities. Many thanks to our favorite language teachers—Renilde, María-José, Nele, Quique, Lieve—it was a lot of fun to learn languages with you. Thanks also to Luk, Leen, Hanne, Cami, Gregory for the short, but enjoyable moments together.

Another big "thank you" goes for the Bulgarian community in Leuven and Brussels! Hyshove, thanks for all the events we shared—the Guitar evenings at Mitko's place, the BBQ and cocktails evenings at Ivo and Stoyan's places, my favorite Domus chicken wings meetings, the national folklore dancing events, and many more! Thanks to all of you—Stefan, Mitko, Ivo, Tsetso, Tsveta, Vihren,

Petar, Vesko and Vesko, Stoyan, Sasho, Daro, Hrisi, Yana, Velina, Tsetso, Vera, Tsveti, Misho, Nadya, Darin, and all the rest of the hyshovete gang.

I am very grateful to all my friends from Bulgaria or other parts of the world who never gave up keeping in touch! And special thanks to the guys who managed to come to visit! Nedi and Borko, Cani and Valya, Bairevi, Kosio, and Maya, Nina, bace Jore, and bace Sime, kako Zo, I am sure we made some unforgettable memories here! Thanks also to all of you who were always finding time to see us when we were in Bulgaria—the whole volleyball gang (Kosio, Maya, Yana, Evgeni, Milenka, Ani, Tacka, Vesko, Leya, Assia, Boiko, Raya, Nina, Mincho, Megi), the eternal high-school friends (Joreto, Yoko, Simo, Niki, Kiro, Barko), the colleagues from the Physics Faculty (Stoyan, Vesko, Gichka, Kiro, Tsetso), the colleagues from Melexis (Nadya, Rosen, Miro, Kolio, Vankata, Simo), g-zha Christakudi.

I also want to thank my family for supporting me, each in their own way, but always with a warm and caring feeling. To my grandparents: thank you so much for being my biggest fans and supporters! To my brother: Bratle, I really enjoyed the philosophical discussions over the chat and above all—having fun on the account of mamichka! I am sure that secretly, she liked it too. To my dad: Tati, thanks for the maybe not so regular, but surely timely support, for your interest in my work and for believing that I can be a good scientist! To my mum: Mamo, I am not sure if I can express how much you helped me during these 4 years—actually, as you have always helped me. Thanks for always adapting to my crazy schedule when I was at home, for always keeping a cozy place for me at home, for always cooking so deliciously and—you know—"for picking up the bits and picking up the pieces."

Finally, I want to thank Stefan. Thank you for… wow, for so many things… It was great to grow up together with you, learning together the sweet and the sour parts of that process. Thank you for being always so reliable and trustworthy. Thank you for all the fun, laugh, and adrenaline we shared. Thank you for never giving up teaching me how to be a better person. You are the best teacher for that, since you are the best man I know. I don't know whether our paths will ever cross again, but I want to wish you success fighting with your own challenges and I want to wish you to find the things which make you happy—you deserve that!

Thank you all!

Best wishes,
Denitza

Contents

1 Introduction 1
   1.1 Surface Plasmon Resonances in Metal Nanoparticles 1
        1.1.1 The Underlying Physics 1
        1.1.2 Plasmonics: Application Highlights 4
        1.1.3 Optical Characterization of Plasmonic Devices 6
   1.2 Scanning Near-Field Optical Microscopy (SNOM) 8
        1.2.1 The Diffraction Limit 8
        1.2.2 Principles of Sub-wavelength Optical Microscopy 15
        1.2.3 Basic Configuration and Practical Realizations 16
        1.2.4 Artifacts and Challenges 23
   1.3 SNOM for Imaging the Magnetic Field of Light 24
        1.3.1 Imaging the Electromagnetic Field Components
               by SNOM 25
        1.3.2 Fundamental Challenges in Imaging the Magnetic
               Field of Light 26
        1.3.3 Experimental Setup for Imaging the Magnetic Field
               of Light 28
   References 29

**2 Imaging the Magnetic Near-Field of Plasmon Modes
in Bar Antennas** 35
   2.1 Introduction 35
   2.2 Results and Discussion 37
        2.2.1 Individual Probe and Sample Characterization 38
        2.2.2 Probe-Sample Coupling: Imaging of the $|H_y|^2$
               Near-Field Distribution of an $l=3$ Plasmon
               Mode in a Gold Bar 39
        2.2.3 Imaging of the $|H_y|^2$ Near-Field Distribution
               of Different Plasmon Modes in a Gold Bar 43
        2.2.4 Plasmon Dispersion Relation Obtained by SNOM 43

	2.3	Conclusions	46
	2.4	Methods	47
		2.4.1 Sample Fabrication	47
		2.4.2 FDTD Simulations	47
		2.4.3 Electric and Magnetic Field Profiles at the Apex of the Probe	48
	References		49
3	**A Near-Field Aperture-Probe as an Optical Magnetic Source and Detector**		53
	3.1	Introduction	53
	3.2	Results and Discussion	54
		3.2.1 Hollow-Pyramid Aperture Probe as a tangential H_y Dipole Source: Intuitive Physical Justification	54
		3.2.2 Correspondence Between the Fields of the Hollow-Pyramid Probe and a tangential H_y Dipole: Simulations	57
		3.2.3 Scanning of the Probe Over a Sample	58
		3.2.4 Experimental Evidence for Equivalence of Collection and Illumination Mode SNOM	60
	3.3	Conclusions	61
	References		62
4	**Magnetic Near-Field Imaging of Increasingly Complex Plasmonic Antennas**		63
	4.1	Introduction	63
	4.2	Results and Discussion	65
		4.2.1 Simple Antennas	66
		4.2.2 Complex Antennas Consisting of Assembled Bars	68
	4.3	Conclusions	72
	4.4	Methods	73
		4.4.1 Sample Preparation	73
		4.4.2 Simulations	73
	4.5	Supporting Information	73
		4.5.1 Simulation of the $|H_y|^2$ Near-Field Map of an Elementary Horizontal Bar	75
		4.5.2 Simulation of the $|H_y|^2$ Near-Field Map of an Elementary Vertical Bar	75
		4.5.3 Simulation of the $|H_y|^2$ Near-Field Map of Complex Antennas	75
		4.5.4 Electromagnetic Field Components of a Representative Complex Antenna	77
	References		78

5 Conclusions and Outlook 81
5.1 Conclusions 81
5.2 Outlook 83

Curriculum Vitae 85

Abbreviations

AFM	Atomic force microscopy
AOTF	Acousto-optic tunable filter
CNT	Carbon nanotubes
EBL	Electron beam lithography
FDTD	Finite-difference time-domain
FTIR	Fourier-transform infrared spectrometer
LDOS	Local density of optical states
PALM	Photo-activated localization microscopy
PMMA	Poly (methyl methacrylate)
SCWL	Supercontinuum white light
SEM	Scanning electron microscopy
SERS	Surface-enhanced Raman scattering
SHG	Second harmonic generation
SNOM	Scanning near-field optical microscope
SPR	Surface plasmon resonance
STED	Stimulated emission depletion
STORM	Stochastic reconstruction microscopy
TEM	Transmission electron microscopy
TIRF	Total internal reflection microscopy
TOA	Tip on aperture
TPL	Two-photon luminescence

Symbols

a	Aperture diameter
A	Lens size
α	Fine-structure constant
$B = \mu_0 H$	Magnetic induction
c	Speed of light
δ_x	Spatial resolution
δ_θ	Angular resolution
D	Distance between lens and sample
e	Elementary charge
E	Electric field
ε_O	Permittivity of free space
F_b	Magnetic force
F_e	Electric force
G	Gap size between two parallel bars
h	Bar height
H	Magnetic field
$\hbar$	Reduced Planck constant
J	Current density
k	Wave number
l	Plasmon mode number
L	Bar length
λ	Wavelength
m	Mass of a moving charged particle
μ_0	Permeability of free space
n	Refractive index
r	Distance between an aperture and a point of observation
r_A	Radius of the first Airy disk
r_B	Radius of an electron's orbit
R	Distance between an aperture and a screen

θ	Acceptance angle of an objective
q	Electric charge
v	Speed of a moving charged particle
W	Bar width

Chapter 1
Introduction

Abstract In this chapter, we first describe the interaction of light with metallic nanostructures and the consequently induced plasmonic effects in the structures. The most important characteristics of such interaction and their significance for various practical applications are discussed. Then, we point out the difficulties related to the experimental characterization of plasmonic effects. In the second section we explain how the scanning near-field optical microscope (SNOM) can overcome those difficulties. The practical experimental realization of such microscope and different variations of the setup are further discussed. In the last section of the chapter we focus on the main subject of this thesis—the imaging of the magnetic field of light. We point out the necessity and the main difficulties related to such measurements. Then, we describe the experimental setup which we use for this purpose.

1.1 Surface Plasmon Resonances in Metal Nanoparticles

1.1.1 The Underlying Physics

When a metal is illuminated with light under certain conditions, for example under total internal reflection, under an angle or in the presence of corrugations in the metal, the free charge carriers in the metal can be driven to oscillate by the incoming electromagnetic field of the light [1]. These oscillations are called surface plasmon polaritons. They are propagating along the metal-dielectric interface and they are evanescently confined in the perpendicular direction. The metal can be nanostructured, so that it forms a cavity for these collective charge oscillation waves. When

---

The results presented in this section are based on and reproduced with permission from:
D. Denkova, N. Verellen, A.V. Silhanek, V.K. Valev, P. Van Dorpe, V.V. Moshchalkov
Mapping magnetic near-field distributions of plasmonic nanoantennas
ACS Nano **7**, 3168 (2013).
Copyright © 2013, American Chemical Society.

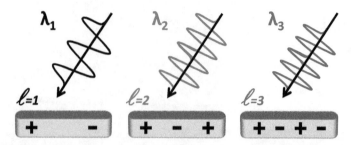

Fig. 1.1 Illumination of a gold bar with light results in excitation of oscillation waves of the free electron gas in the metal—plasmons. At different wavelengths, different resonance modes are excited – $l = 1$, $l = 2$ and $l = 3$ modes are shown here. The distribution of the charges for each mode is schematically indicated

the plasmon waves are confined in such cavities, at certain wavelengths, the plasmon waves fit in the nanocavity to form standing-wave patterns, similar to resonant modes of a guitar string. The spectral position of the resonances depends on the composition, shape and size of the nanoparticle and on the local dielectric environment.

An illustrative sketch of these plasmonic effects is presented in Fig. 1.1. The metal nanoparticle playing the role of a plasmonic resonator is a gold bar. When this bar is illuminated with light polarized along its long axis, at the resonant wavelengths, standing-wave like Fabry-Pérot resonances, also called surface plasmon resonances (SPR) are excited. The charge distribution along the bar for the different resonance modes is sketched in the figure. The resonance modes are typically indexed by the number of half-plasmon wavelengths fitting the antennas' cavity, respectively the mode number corresponds to the number of charge maxima minus one. The $l = 1$, $l = 2$ and $l = 3$ resonance modes are shown in the figure.

It should be noted, that due to symmetry considerations, the excitation of the even plasmon modes is forbidden under plane wave illumination at normal incidence [1]. To excite those modes, a certain retardation should be introduced in the field illuminating the particle. This can be realized in different illumination configurations, for example by illumination at an angle (as illustrated in Fig. 1.1) or by illumination with a dipole source (Fig. 1.2).

To further discuss the plasmonic effects, we will focus on a specific plasmonic nanoantenna, representative for the structures studied in this thesis. For simplicity, we will consider only one plasmon mode, for example the $l = 3$ mode—Fig. 1.2. The specific metal nanoparticle is again a plasmonic bar with the following dimensions: length $L = 1120$ nm, width $W = 70$ nm and height $h = 50$ nm—Fig. 1.2a. The particle is covered by a 30 nm glass layer. The necessity for such a layer is required for the experiments and will be discussed later. For now, it is sufficient to note that besides introducing a small resonance shift, the layer does not fundamentally affect the plasmonic effects in the metal.

To simulate the various characteristics of the excited plasmons in such a bar, we use illumination by an electric dipole source (indicated with a black star in Fig. 1.2a), polarized along the bar axis. The result from the rigorous simulation of

1.1 Surface Plasmon Resonances in Metal Nanoparticles

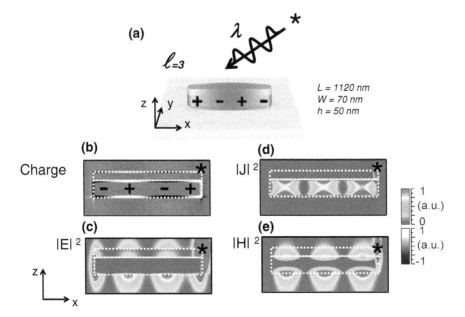

Fig. 1.2 **a** The $l = 3$ plasmon mode in a gold bar can be excited by a dipole source illumination (*black star*). **b** Four maxima are observed in the simulated charge and **c** electric field profile. **d** At the complementary positions, the corresponding current density and **e** magnetic field distributions exhibit three maxima

the charge distribution at wavelength $\lambda = 1270$ nm is shown in Fig. 1.2b. Indeed, the four charge anti-nodes for the $l = 3$ plasmon mode can be clearly observed. The disturbed charge equilibrium results in the concentration of the electric field in four respective maxima Fig. 1.2c. The enhancement of the current is occurring complementary to the charge maxima and three positions with high current density can be identified in the simulated image—Fig. 1.2d. Accordingly, these positions determine the positions of magnetic field localization in the bar—Fig. 1.2e.

One of the most important consequences of the interaction of light with metal is the fact that the light gets highly localized in the near-field of the metal. A more precise definition of the near-field and the effects occurring there will be given in Sect. 1.2.1. In brief, the near-fields are localized in a region roughly one wavelength away from the surface. They are hardly or not at all radiating towards the far-field.

In a sense, we can perceive the plasmonic nanoparticles as a kind of optical antennas—devices which convert the freely propagating light waves into localized energy and the other way around [2]. From one side, this is in the basics of most of the plasmonic applications—the confinement of the light in nanoscale volumes can result in significant enhancement of the local electromagnetic field in the vicinity of the structures. In the rest of this section, we highlight some of the interesting plasmonic applications. From the other side, the very same properties from which we benefit in the applications, namely the field localization in the close proximity of the sample, make the characterization of the plasmonic effects very difficult, as described in the last part of this section.

1.1.2 Plasmonics: Application Highlights

Most of the plasmonic applications are based on the localization of the light in nanoscale dimensions, which can result in very high field enhancements in the near-field of metallic structures. For specific cases, enhancements up to 10^6 times have been reported in the gaps between metallic nanoparticles. This can be extremely beneficial for increasing the interaction of light with, for example, molecules positioned in those, so called, hot spots. The high fields can also enhance typically weak effects, such as non-linear interactions for example. Additionally, the fact that the field is mainly confined in the very close proximity of the nanoparticle, means that we can use with great precision these fields as, for example, very local nano-sized sources of light or heat.

In this section we highlight several important application of plasmonics and discuss on the crucial parameters of the light matter interaction, which need to be characterized and tuned to optimize the respective applications.

Bio- and Chemical-Sensing

As mentioned above, the plasmon resonances are extremely sensitive to the refractive index of the local dielectric environment of the metal. This effect has been broadly explored for sensing applications: the metal can be functionalized, so that an analyte is attached to it. Then, the resulting change in the plasmon resonance condition due to the presence of the analyte can be detected as wavelength, intensity or angle shifts [3–5].

It has been shown that the surface plasmon effects, namely the strong electric field enhancement in the vicinity of the metal surface, can result in very strong enhancement of the Raman scattering effect—so called Surface Enhanced Raman Scattering (SERS) [6, 7]. Enhancement up to 10^{11} of the Raman signal have been reported, which is sufficient for studying analytes even at single molecule level [8, 9].

Thus, to achieve well-defined hot-spots with high enhancement factors and high refractive index sensitivities for sensing applications, it is crucial to be able to accurately measure the plasmon resonance wavelengths and to precisely map the electric field enhancements on the metal surface.

All-Optical Chips

The current state-of-the-art computing devices are based on *electronic* circuits, which operate by controlling the transport and storage of electrons. Two of the main components of the digital circuits are (i) the individual logic elements (transistors), which perform the actual computational process and (ii) the interconnects (typically based on copper wires), which transport the digital information between the different parts of a microprocessor. The advances in the Si technology allow continuous miniaturization of the electronic components and it is nowadays a routine to produce fast transistors with channel dimensions on the order of tens of nanometers. However, while downscaling the dimensions can improve the performance of the transistors, it worsens the performance of the interconnects [10]. Reducing the width of the interconnects leads to increase of the signal delay [10] and this is one of the main factors limiting nowadays the computational speed.

In contrast, *optical* communication systems, based on optical fibers and photonic circuits can transport digital data with a capacity more than 1000 times higher than that of the electronic interconnects [11]. However, the optical elements have the disadvantage of being about 1000 times bigger. This is because the wavelength of the light used in photonic circuits is on the order of 1000 nm, and the optical components cannot have dimensions close and below the wavelength of the light, due to the appearance of diffraction effects.

Consequently, an ideal configuration would be to have a circuit with nanoscale dimensions which can carry both optical signals and electrical currents. However, integrating electric and photonic circuits is highly limited due to the mismatch of their respective sizes. A solution to this problem might be the field of plasmonics—the confinement of the light in nanoscale dimensions around metallic nanostructures offers both the capacity of photonics and the miniaturization of electronics [10–12]. Therefore, a lot of efforts are concentrated currently in overcoming the intrinsic losses of plasmonic devices and the development of plasmonic chips, with all necessary components [13]—sources, detectors, waveguides, switches [14], modulators, etc.

Metamaterials

Literally, in Greek, the word "meta" means "beyond". Thus, the term "metamaterials" refers to "beyond" conventional materials. Those are typically man-made materials, designed to have properties, not found in nature. For example, they can exhibit artificial magnetism [15, 16] or negative refractive index [16–20]. The latter means that when the light is passing through such a material, it refracts in the opposite direction to normal materials. This requires a re-examination of many optical phenomena—for example the Doppler shift is reversed, as well as the direction of the Cherenkov radiation [17]. The negative refractive index should allow the construction of a "perfect lens"—a flat piece of material, which can provide a "perfect" image of an object, with resolution well below the wavelength of the light [21]. Even magic-like objects and properties have been proposed, such as an invisibility cloak, able to "hide" objects from the observer [22–24]. Presently, the field is naturally evolving towards the construction of metadevices—devices, based on metamaterials, allowing extraordinary functionalities, including switching, tunability, non-linear effects, sensing, etc. [25].

The functionality of these materials is based on the fact that they are constructed from artificial, man-made units—"meta-atoms", which have sub-wavelength dimensions. Since those "meta-atoms" are man-made, their structure, properties and interaction can be freely tailored, the only limitation being the imagination of the scientist and the available fabrication techniques. This opens possibilities to realize physical phenomena, which are difficult or impossible to achieve with conventional materials.

One of the main reasons allowing the metamaterials to achieve this broad spectrum of extraordinary properties is the following: the light, as an electromagnetic wave, consists of oscillating electric and magnetic fields, interconnected via Maxwell's equations. However, at optical frequencies, the interaction of atoms with the magnetic field of light is normally weak and thus, the light-matter interaction is mainly governed by the electric field of light [26, 27]. This is further elaborated in the last section of this chapter. One of the advantages of constructing artificial "meta-atoms" is that they can be designed, so that they interact also with the magnetic field of light, thus opening a whole new world of possible functionalities.

Therefore, only recently, with the flourishing of the field of metamaterials, the exploration of light-matter interactions mediated by the magnetic component regained a lot of interest. During the last years the optics community has invested tremendous efforts in this direction of enhancing and characterizing the magnetic light-matter interactions [28–32]. The presented work in this thesis contributes to this search for magnetic field imaging techniques.

1.1.3 Optical Characterization of Plasmonic Devices

To be able to optimize the various plasmonic applications described above, we need first to be able to characterize the plasmonic effects. Since we are interested in studying the interaction of light with metals, optical microscopy would seem to be the technique of choice. However, due to the nature of this interaction, certain difficulties arise, which are discussed below. Actually, the very same properties from which we benefit in the applications, make the characterization of the plasmonic effects very difficult. In this section, we elaborate on the challenges in the characterization of plasmonic devices, with focus on optical characterization techniques.

Nanoscale Resolution

The first problem regarding the study of plasmonic devices is a general limitation, which originates from the fact that, since the plasmonic effects localize the light in nanoscale volumes, characterization of those effects requires nanoscale resolution. There are different techniques which can provide structural and topographical information with nanoresolution, such as Atomic Force Microscopy (AFM), Scanning Electron Microscopy (SEM) and Transmission Electron Microscopy (TEM). However, in terms of optical characterization we are very limited, because the conventional optical microscopes operate in the far field region and their resolution is fundamentally limited by diffraction effects to roughly about half of the wavelength [33]. The origin of this, so-called, diffraction limit will be elaborated in Sect. 1.2.1. Essentially, it means that the microscopes operating with visible light do not have sufficient resolution to study nanomaterials.

Still, conventional light microscopy provides specific complementary information which cannot be obtained with the other mentioned high-resolution techniques. An optical microscope works with well-known mechanisms such as reflection, transmission, absorption, scattering, which provide readily interpretable information and direct optical characterization. Spectroscopy allows characterization of different energy levels in semiconductors, structural and chemical specific information and gives information about resonances of various plasmonic nanoparticles. Biological samples could not always withstand high vacuum conditions, needed for example in the SEM and TEM, so optical microscopy is a necessity for investigating such samples. Additionally, light microscopy has the advantage of being relatively non-destructive and cheap and does not always require specific operation conditions.

Therefore, achievement of nano-resolution optical microscopy turned out to be crucial for all fields of modern nano-science. Although, for over a century it has been thought that optical microscopy is fundamentally limited by diffraction, recently various ways to circumvent the diffraction limit and achieve sub-wavelength resolution optical microscopy have been proposed. Such are for example: fluorescence microscopy, total internal reflection microscopy (TIRF), two-photon excitation microscopy, stimulated emission depletion (STED) microscopy, photo-activated localization microscopy (PALM), stochastic reconstruction microscopy (STORM). However, most of those techniques have quite specific requirements for the type of samples which can be imaged, require labelling of the samples or complicated instrumentation and data processing. The principles of those techniques and comparison between them can be found in different reviews on the subject [34–39].

Access to the Near-Field

The second difficulty for studying plasmonic effects is specific for this kind of effects: since they have the ability to concentrate the light in the close vicinity of the metal surface, the characterization technique needs to be able to access the near-field of the metal. This means, that it has to be able to somehow access fields, which are only present to a few hundreds of nanometers from the particle. Typically, some kind of a scatterer of those fields is needed to make them accessible to detectors, positioned further away from the structures, *i.e.* in the far-field.

Nowadays, one of the most powerful tools for sub-wavelength imaging is the scanning near-field optical microscopy (SNOM) technique [40–44]. In addition to the tens of nanometers resolution capability, the technique can image the near-field of the sample, which is, as explained above, a requirement for studying plasmonic structures. Therefore, we have chosen to use and develop this technique for characterization of plasmonic structures. In the next section we will discuss the physical idea behind the technique, the practical realization and the remaining challenges to its usage and applicability.

1.2 Scanning Near-Field Optical Microscopy (SNOM)

In the search for optical techniques capable of imaging the near-field of light with nano scale resolution, the SNOM has become one of the most popular techniques. In this method, by positioning a probe in the evanescent field of the sample, the near-field can be imaged with resolution, determined by the size of the probe and the probe-sample distance but not by the wavelength as in the conventional optical microscopy. Typically, the resolution is in the range of tens of nanometers, although sub-10 nm resolution has been reported [45, 46]. Additionally, in contrast to conventional microscopy, where only information about the light intensity is typically obtained, the SNOM can separately access the different components of the electromagnetic field. Which component will be picked up depends on the type of probe and measurement configuration.

In this section we will first elaborate on the origin of the limitations of the standard optical microscopy, more specifically—on the origin of the diffraction limit. Then, we will explain what is the physical idea behind their circumvention with the SNOM and how in practice such microscopes are realized. Since the type of information obtainable by SNOM is mainly determined by the used probe type and configuration, we will further summarize the most popular probes and SNOM configurations. The above mentioned advantages of the method come at a certain price—the presence of the probe in the near-field of the sample disturbs that field, so the influence of the probe on the measured field has to be carefully considered and taken into account. The main challenges in the SNOM technique are discussed in the last part of this section.

1.2.1 The Diffraction Limit

We classically think of light as rays, travelling in straight lines. However, when the light passes near a barrier or through an aperture, it tends to bend and spread out. This effect is called diffraction and it is more pronounced when the characteristic sizes of the barrier/aperture are comparable or smaller than the wavelength of the light. The diffraction limit is the fundamental limitation of the resolution of optical imaging systems, due to the diffraction of light. The concept of the diffraction limit was put on a solid ground in 1873 by the German physicist Ernst Abbe [33] and it was further refined by Lord Rayleigh in 1896 [47]. Rigorous calculations and explanations of the diffraction theory of light can be found in different sources [33, 44, 47, 48]. Here, we present an intuitive physical explanation of the diffraction limit.

Transmission of Light Through an Aperture ($a \gg \lambda$): Geometrical Optics Approximation

We will consider a hole in an opaque screen (object) illuminated by a plane wave and observe the image on a distant screen—Fig. 1.3. According to classical ray optics, the

1.2 Scanning Near-Field Optical Microscopy (SNOM)

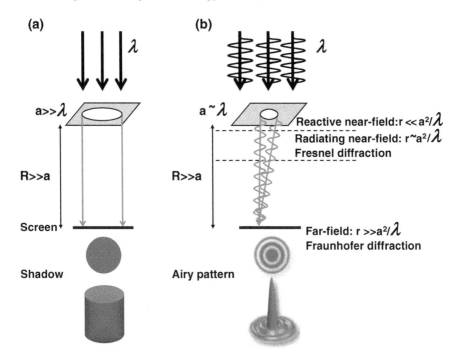

Fig. 1.3 The transmission of light through an aperture in a screen depends on the relative dimensions of the aperture compared to the wavelength of the light and the observation distance (r). **a** The light transmitted through a hole with dimensions $a \gg \lambda$ gives a sharp shadow of the hole on a distant screen. **b** If the dimensions of the hole are $a \sim \lambda$, diffraction effects start to play a role—an Airy pattern of alternating *bright* and *dark spots* is observed on a distant screen. The diffraction pattern in the near-field zone is more complex and depends on the distance to the screen [49]

edges of the hole will cast a sharp shadow and we should see a bright spot with sharp edges on a dark background—in Fig. 1.3a. In the figure, high intensity is represented with blue color, and low intensity—with white. This is indeed what happens if the hole's dimension (a) is much bigger that the wavelength of the light (λ): $a \gg \lambda$ and the screen is positioned at a distance $R \gg \lambda$. This is, most often, the situation in our everyday life—since the wavelength of the light is much shorter than the physical dimensions of the objects around us, we most commonly see sharp objects and shadows around us.

Diffraction of Light Through a Small Aperture ($a \sim \lambda$)

We will now do the same experiment, but shrink the diameter of the hole. The theory of diffraction by small apertures has been developed by Kirchhoff and then further refined by Smythe and others. When the dimensions of the aperture become comparable with the wavelength of the light ($a \sim \lambda$), the wave-like nature of the light becomes important and diffraction effects start to play a role—Fig. 1.3b. According to Huygens' Principle, each point from the aperture may be considered as a source of secondary spherical waves. The wavefronts of these secondary sources start to interfere with each other, giving rise to a much more complicated picture of the transmitted light. The electromagnetic field of the transmitted radiation varies characteristically with the distance from the hole. Although these changes are continuous with distance, typically three distinct spatial regions can be defined with smooth transitions between them—Fig. 1.3b.

Reactive near-field region is the area in the immediate vicinity of the aperture, up to a distance r of around one wavelength ($r < \lambda$). More strict definition, implying the aperture size satisfies: $r \ll a^2/\lambda$. The fields in that region are not radiating and decay with the square or the cube of the distance from the hole [50]. At distances longer than one wavelength those fields are negligibly weak. Thus, basically the information about the object contained in these fields is not reaching further away than about one wavelength from the aperture. The term 'reactive' in the definition of this near-field zone refers to the fact that the fields in this region which are perturbed by the interaction with a present object, will feed-back to the aperture and affect the transmitted radiation.

Radiating near-field region, also often referred to as Fresnel region, occupies the space beyond the reactive near-field region, at distances $r \sim a^2/\lambda$. For the formation of the diffraction pattern in this zone, each point of the diffracting system contributes differently depending on its relative position to the observation point, those being closer playing the most important role. As a result, the near-zone fields are complicated in structure and the diffraction pattern varies with distance from the source. The waves are not plane waves and the relationship between the electric and the magnetic field is complex.

Far-field region, or Fraunhofer region is considered at distances $r \gg a^2/\lambda$. In this zone, the waves can be considered plane waves, thus the electric and the magnetic fields are perpendicular to each other and oscillate in phase. Their magnitudes are related simply by $E = Bc$, where c is the speed of light. For the formation of the diffraction pattern in the far-field zone the whole diffracting system contributes. As mentioned above, each point from the aperture becomes a source of secondary spherical waves (Huygens' Principle). The interference of those waves on a screen positioned in the far-field zone gives rise to a pattern consisting of a bright spot, surrounded by alternating bright and dark rings (Fig. 1.3b), instead of a sharp image of the hole as in the case of a big aperture (Fig. 1.3a). This is called Airy diffraction pattern and it was theoretically described by Airy in 1835 [51]—Fig. 1.3b [49]. The first minima in the pattern appears at position from the center:

$$r_A = 1.22 R \frac{\lambda}{a} \quad (1.1)$$

Actually, for big apertures compared to the wavelength, the effect is also present, but less noticeable.

Diffraction of Light Through a Deep Sub-wavelength Aperture ($a \ll \lambda$)

We will now shrink the hole even further to sub-wavelength dimensions ($a \ll \lambda$). In this case the theoretical models developed by Kirchhoff for the case above do not hold anymore. Bethe was the first one to tackle the problem of light diffraction by a deep sub-wavelength hole in a metal screen [52]. He provides an analytical result for the transmitted field through such an aperture. Bethe's solution for the near-field region was later corrected by Bouwkamp [53].

The physical processes occurring when the light is diffracting through such an aperture are elaborated in Chap. 3, Shortly, in this case the near-fields of the aperture can be treated by static or quasi-static methods. The radiation through the aperture is actually equivalent to the radiation of an in-plane magnetic dipole and an electric dipole normal to the plane. The fields get localized in the vicinity of the hole, exponentially decaying away from it—Fig. 1.4 [54]. The decay length of the near-field evanescent wave is about 20 % of the aperture diameter, thus the aperture becomes a local nano-sized source. The largest portion of the transmitted fields is localized in this nano-sized region and cannot propagate to the far-field. Information from them is lost and cannot contribute to the formation of the image in the far-field.

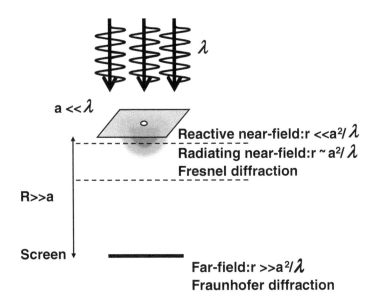

Fig. 1.4 The light diffracted by a deep sub-wavelength aperture is strongly localized in the near-field of the sample and only a very small portion of it propagates to the far-field. The aperture becomes a localized source with nano-sized dimensions. However, in the far-field, the transmitted fields are practically not able to contribute to the formation of an image

Diffraction Limit of Conventional Optical Microscopy

Practically, all conventional microscopes operate in the far-field regime—the illumination and collection objectives are positioned in the far-zone of the sample. Hence, as described above, information concealed in the non-propagating near-field zone can not be obtained by such microscopes. Features with sizes comparable with the wavelength appear in the image with the blurry edges of the Airy patterns instead of sharp shadows. The idea behind the origin of the diffraction limit in a conventional far-field microscope system will be further elaborated below.

For simplicity, we consider a very simplified microscope, consisting of only one lens, playing the role of an objective and focusing the light from an object on a distant screen. As an object, we take a point-like source (P)—a point from the sample plane, scattering the incoming light—Fig. 1.5a. The collection system has a finite size (A) and will collect part of the scattered light (in blue), while part of the light scattered under close to grazing angles will be lost (in red). The collected light is focused by the lens on the screen. Due to the fact that the lens has a finite size, the situation is, in a way, similar to the one described above about the circular aperture. In this case the lens is playing the role of the aperture, in a sense that only the rays reaching the lens will be collected and used for the construction of the image. Thus, analogously,

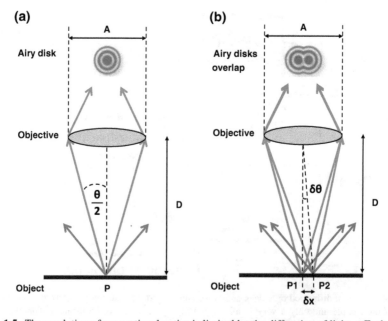

Fig. 1.5 The resolution of conventional optics is limited by the diffraction of light. **a** Each point from the sample is appearing in the image as an Airy pattern. **b** The Airy patterns of two closely positioned object points overlap. According to the Rayleigh criteria, the smallest distance at which the two points are still observed as separate objects is $\delta x = 0.61 \, \lambda/NA$. This is the resolution of a far-field optical system. The Airy disks of points positioned at closer distances are overlapping, thus such points are not distinguishable

1.2 Scanning Near-Field Optical Microscopy (SNOM)

the point source appears on the screen not as a point, but as an Airy disk pattern with a finite size, determined by the wavelength and the size of the objective. Basically, every point from the object will appear in the image as an Airy disk.

Naturally, the next question which arises is: what is the resolution of such a system, meaning what is the minimum distance between two points at which we still see them separate from each other. If we try to see two points which are close together, their Airy disks will start to overlap. The points are still distinguishable if the central Airy disk of one of them coincides with the first minimum of the Airy pattern of the other—so called Rayleigh criteria (Fig. 1.5b). If we get the points closer, they are not distinguishable anymore. According to this criterion, the minimum angular separation possible for a far-field optical system is:

$$\delta\theta = 1.22 \frac{\lambda}{A} \quad (1.2)$$

Here, we have used the small angle approximation: $\sin\theta \approx \tan\theta \approx \theta$. Still within this approximation, from Fig. 1.5b we can estimate that:

$$\delta\theta = \frac{\delta x}{D} \quad (1.3)$$

where δx is the distance between the two points P1 and P2 and D is the distance between the lens and the sample. Thus, combining Eqs. 1.2 and 1.3 we obtain:

$$\delta x = 1.22 \frac{\lambda D}{A} \quad (1.4)$$

Here, it is convenient to introduce the concept of a numerical aperture NA of the lens. It is a measure of the maximum acceptance angle of the lens and it is defined by the refractive index of the imaging medium (n; usually air, water, glycerin, or oil), multiplied by the sine of the aperture angle ($\sin\theta$).

$$NA = n \sin\frac{\theta}{2} \quad (1.5)$$

Again, the geometry of the system in Fig. 1.5b, allows us to consider small angle, thus we can approximate that:

$$\sin\frac{\theta}{2} \approx \frac{\theta}{2} \approx \frac{A/2}{D} = \frac{A}{2D} \quad (1.6)$$

Hence, the resolution of a far-field optical system according to the Rayleigh criteria can be expressed in the more commonly known form:

$$\delta x = 0.61 \frac{\lambda n}{NA} \quad (1.7)$$

Thus, it is the diffraction effects which limit the resolution of standard far-field optical microscopes. Every point from an image is in fact an Airy disk and we can not resolve two points which are closer than the distance δx defined above.

According to formula 1.7, the resolution of such a system can be increased either by using shorter wavelengths or by increasing the *NA*. Usage of shorter wavelengths is, however, restricted by the limited sources and, in general, optical elements in the UV. Most of the optical components become highly absorbing in the UV and exotic and expensive materials are required for production of optical components suitable for that wavelength region. Also, this implies the limitation that the wavelength of the light used for imaging is dictated by the desired resolution and not by the spectral properties of the sample.

The *NA* can be increased to certain amount by using immersion oil objectives, but still the best achievable *NA* is about 1.6. Actually, it is the finite size of the collection system which prevents us from obtaining a point like image on the screen [44]. Since some of the rays coming from the object are not collected by the optics, the lack of this information prevents the full reconstruction of the image. Collecting a bigger portion of the rays coming from the object results in shrinking of the Airy disk size, and allows obtaining an image closer to the actual point object. The shrinking of the Airy disk size, and respectively improvement of the resolution, when increasing the portion of the collected light by the objective is illustrated in Fig. 1.6.

Practically, in the best case, the resolution of far-field optical microscopy is limited to about 200 nm.

Having described the origin of the diffraction limit, we will further discuss how this limit can be overcome.

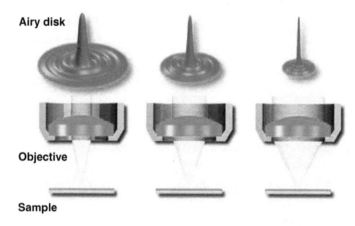

Fig. 1.6 Increasing of the *NA* of an objective leads to shrinking of the Airy disk size and respectively, improvement of the resolution [49]

1.2.2 Principles of Sub-wavelength Optical Microscopy

As described above, the basics of the diffraction limit of conventional far-field microscopes is the fact that some of the light scattered from sub-wavelength features can not reach the detection system because (i) it is localized in the near-field and does not propagate to the far-field and/or (ii) the finite dimensions of the collection system do not allow collection of light scattered at all angles. The lack of this information prevents the full reconstruction of the image. This loss of information occurs when the illumination and the collection are realized in the far field. These are the typical conditions in which all conventional optical microscopes operate—they use objectives to provide far-field illumination and detection. However, if a microscope which allows breaking one of those conditions is built, an opportunity to circumvent the diffraction limit appears.

It was in 1928 when Synge first proposed a way to circumvent the diffraction limit and achieve sub-wavelength resolution with an optical microscope [40]. Synge's idea was conceptually quite simple—to illuminate the sample by a light source with sub-wavelength dimensions, positioned in the near-field of the surface—Fig. 1.7. To realize the idea in practice, he proposed forming an aperture in an opaque metal screen with dimensions much smaller than the wavelength and illuminating the sample through that aperture. This illumination is equivalent to the diffraction of light through a deep sub-wavelength aperture explained in the previous section and elaborated in Chap. 3. Shortly, the hole becomes a *nano-sized* source, whose fields are *localized* in the near-zone of the aperture and practically do not propagate to the far-field.

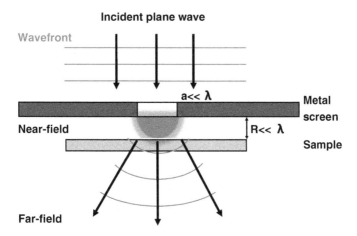

Fig. 1.7 The original idea for construction of a sub-wavelength resolution microscope (Synge [40]) was to illuminate a sample with the evanescent field of a sub-wavelength aperture in a metal screen. For this purpose, the sample has to be positioned in the near-field of the aperture and raster scanned with it. At each scanning position the near-field gets scattered to the far-field by the sample features. Recording of the scattered light results in sub-wavelength resolution image, determined by the size of the aperture and the distance between the probe and the aperture

We can now use this source for local illumination of a sample, if the sample is brought in the near-zone of the source. This will result in illumination of only a restricted part of the object with sub-wavelength dimensions determined by the size of the source and the distance between the source and the sample. The illuminated part of the sample will scatter the near-field of the source towards the far-field, where the light can be collected by a conventional far-field collection objective. The critical moment is, that the scattering is occurring only from the illuminated part of the sample, which is with sub-wavelength dimensions. Thus, the scattered and collected far-field will carry information coming from a sub-wavelength area, which means that we can achieve a sub-wavelength resolution. Scanning the source and recording the transmitted/scattered light can produce a sub-wavelength resolution optical image.

The achievement of sub-wavelength resolution can be also realized in the inverse configuration scheme, where the illumination is performed via a conventional far-field objective, whereas the sub-wavelength hole is used as a local light detector. Since the "detector" is positioned in the near-field, it can collect the evanescent fields which are otherwise not reaching the far field detectors, preventing them to make a complete reconstruction of the object.

It should be noted, that such a near-field microscope allows not only better optical resolution, but it allows also visualization of the near-field of the sample. Therefore, the combined capability of near-field imaging and nanoresolution makes the SNOM one of the most popular techniques for studying plasmonic effects.

Although quite straightforward, at that time Synge's idea was difficult to implement from a technical point of view—the tiny aperture formation and maintenance of nanometers sample-aperture distance were problematic. The first experimental demonstration of this idea was made in 1972 by Ash and Nicholls using microwave radiation ($\lambda = 3$ cm) and an aperture of 0.5 mm [41]. In 1984 the groups of Pohl at IBM Zurich and Lewis at Cornell University (USA) were the first to report sub-wavelength optical measurement in the visible range [42, 55]. Since then, numerous variations of this idea for diffraction free optical microscopy have been developed and implemented for various studies. Depending on the specific sample and characteristics which need to be studied different types of SNOM configurations, feedback mechanisms and near field probes, might be used. Those experimental realizations will be discussed in the next section.

1.2.3 Basic Configuration and Practical Realizations

Different SNOM configurations, near-field probes, detection and excitation schemes have been introduced for miscellaneous applications. In this section we will first describe a general SNOM setup and its most popular variations. Next, we will discuss on the most used types of probes—aperture probes, apertureless ones and some specifically developed probes.

A schematics of a typical SNOM realization is presented in Fig. 1.8. The light from a laser is delivered to a metal coated tapered optical fiber, used for sample

1.2 Scanning Near-Field Optical Microscopy (SNOM)

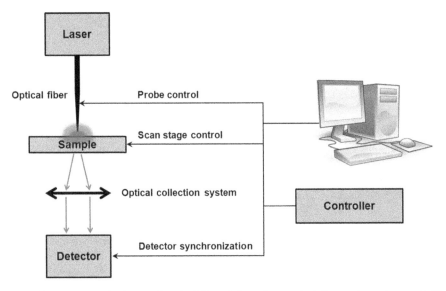

Fig. 1.8 In a typical SNOM the light is delivered through an optical fiber probe to the near-field of the sample. The scattered or transmitted light is recorded synchronously with the scanning of the probe. This synchronization and the maintenance of a nanometers probe-sample distance are controlled by a computer and a controller

illumination. The sample is positioned on an x-y stage and the fiber is maintained at a fixed nanometer distance above the sample by a feedback mechanism. Usually, the sample is raster scanned to obtain an image and the illumination and collection systems are fixed. This avoids complications in the optical path alignment during the scan, but the reverse configuration is also used. The scattered near-field signal is collected in transmission and sent to the detector. Filters, polarizers and other optical elements could be added, if needed. The system is synchronized by a controller and a computer.

Depending on the sample and the information needed to be extracted, different variations of that basic SNOM configuration have been introduced. The most popular ones are summarized and illustrated in Fig. 1.9.

(a) *Transmission illumination mode*—illumination through the SNOM tip in the near-field and collection by an objective in the far-field.
(b) *Transmission collection mode*—illumination from the far-field and collection of the near-field with the SNOM tip. As an option, the far-field bottom illumination might be fixed to the sample stage and they can be scanned together under the tip—necessary for waveguide studies, for example.
(c) *Reflection illumination mode*—illumination through the SNOM tip and collection of the far field by a detector positioned on the side.
(d) *Reflection collection mode*—far-field illumination of the sample from the side and collection of the near–field with the SNOM tip.

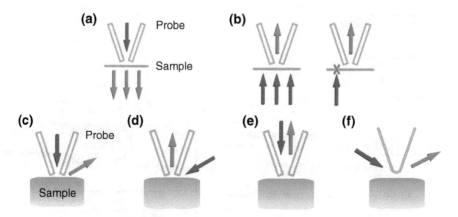

Fig. 1.9 The most popular SNOM configurations are: **a** Transmission illumination mode. **b** Transmission collection mode. **c** Reflection illumination mode. **d** Reflection collection mode. **e** Illumination collection mode. **f** Apertureless scattering SNOM

(e) *Illumination collection mode*—the same aperture tip is used for illumination and collection.
(f) *Apertureless scattering SNOM*—the sample is illuminated from the side by a far-field objective. The local field is scattered by the SNOM tip and the scattered signal is collected again by a far-field detector positioned on the side.

The apertureless mode is widely used, because it can provide higher resolution than the apertures modes [45, 46]. However, it comes at the price of complex extraction of the useful signal from a significant far-field background. From the aperture SNOM modes, one of the most popular ones is the transmission illumination mode. It is relatively easier to align and cleaner of far-field background compared to the other modes. It has been shown that the reciprocity theorem holds also for the near-field, meaning that images made in illumination and in collection mode should be identical [56–58]. The choice of SNOM configuration and probe depends on the specific application of the setup, as will be explained in the last section of this chapter.

As already mentioned above, one of the most critical points in the SNOM instrumentation are the near-field probe and the control of the sample-probe distance. Below we describe the most widely used SNOM probes with their feedback mechanisms and some specially developed state-of-the-art probes with specific purposes.

Aperture Probes

Tapered optical fiber probes.
The most commonly used SNOM probes are tapered optical fiber probes. Those are optical fibers, pulled and thinned at the end until a sub-wavelength diameter is reached. The glass is typically coated with metal to prevent the light from leaking out of the fiber walls. Usually, Al is used for coating due to its small skin depth—the light penetrates until about 13 nm at 500 nm wavelength. A typical optical fiber probe is

1.2 Scanning Near-Field Optical Microscopy (SNOM)

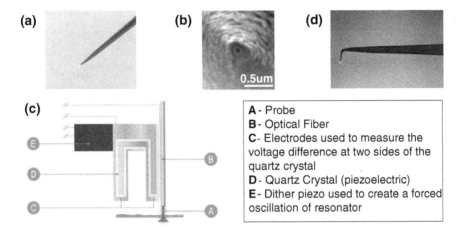

Fig. 1.10 Optical fiber probes are one of the most popular SNOM probes. **a** The light coming through the fiber is also visible in a far-field microscopy image [59]. **b** SEM image of a typical optical fiber probe [60]. **c** The probe-sample distance is controlled via a shear-force feedback [61]. **d** In the case of bent optical fiber probes, cantilevered feedback mechanism is possible [62]

shown in Fig. 1.10a [59]—some of the light, propagating through the fiber and being scattered to the far-field is clearly visible. A close up SEM view of the tip is also included—Fig. 1.10b [60].

These probes require the so called shear-force feedback method to control the probe-sample distance. The fiber is attached to one of the shoulders of a tuning fork, which is oscillated at its resonance frequency in lateral direction—Fig. 1.10c [61]. When the probe approaches the sample, different interaction forces appear which lead to damping of the oscillation at one of the shoulders of the tuning fork and appearance of a voltage difference between the two electrodes C. This voltage difference is monitored and used as a feedback signal to maintain fixed sample-probe distance.

The tapered optical fiber probes have good transmission characteristics and are relatively easy to produce. It is however very difficult to achieve mass production with well reproducible properties of the probes. The main drawback of the optical fiber probes is their fragility. The probes are quite stiff and if unwanted contact occurs between the probe and the sample, both of them can be easily damaged.

There exist also bent optical fiber probes, which allow feedback control of the probe-sample separation similar to conventional cantilevered AFM, as explained in the next section—Fig. 1.10d [62]. There is less chance of damaging the probe and the sample during the scan, but these bent probes are difficult to fabricate and a lot of signal losses occur at the bend. This requires increasing of the light intensity, which might lead to heating of the probe and the sample and to more stray light background, contaminating the useful signal.

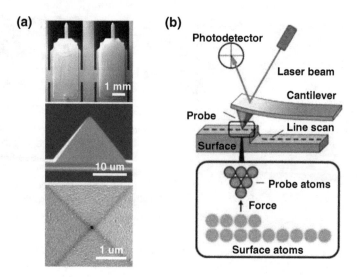

Fig. 1.11 **a** SEM images of the hollow-pyramid SNOM probes [63]. **b** The cantilever-based feedback mechanism allows more accurate control of the probe-sample separation than the shear-force feedback used for the optical fiber probes [64]

The tapered optical fiber probes offer a lot of flexibility in the geometrical configuration of the SNOM setup, since the probe can be made a stand-alone piece of equipment, not integrated in the microscope. Thus, the probes themselves can be scanned without introducing serious shift in the optical path. Additionally, more than one probe can be included in a measurement with position and time synchronization between them and/or the sample scan stage.

Cantilever based probes.
Recently, a novel type of probes has been introduced—cantilever based SNOM probes, analogous to a hollow AFM tip. They consist of a hollow SiO_2 pyramid, coated with metal and mounted on an elastic cantilever arm. Different SEM views of such cantilevered probe are shown in Fig. 1.11a [63].

The probes are produced on a silicon wafer, by the standard, well developed methods of the silicon industry. This allows batch fabrication with well defined and reproducible properties. These probes are very robust and relatively easy to use. Additionally, they maintain relatively well the initial polarization of the transmitted light [65]. This is not the case for the optical fiber probes, where bending of the fiber or strain induced effects can easily change the polarization state.

Integrating a SNOM tip on an AFM cantilever allows imaging of the sample in all well developed force-microscopy modes (contact mode, tapping mode, non-contact mode), thus offering a lot of options for probe-sample distance control. The most popular method for feedback control of cantilever based probes is the beam deflection laser feedback, illustrated in Fig. 1.11b [64]. A laser beam is reflected by the back side of the cantilever and sent to a segmented photodetector.

1.2 Scanning Near-Field Optical Microscopy (SNOM)

In contact mode, the cantilever is brought close to the surface and dragged in lateral direction. When the cantilever is in close proximity of the surface, different forces are causing it to bend (main contribution is usually coming from Van der Waals forces). During the scan the probe-sample interaction forces change, which respectively causes changes in the bending of the cantilever. Consequently, the position of the reflected laser spot on the photodetector is changing. The feedback loop is monitoring these changes in order to keep the desired probe-sample distance constant.

This method is very straightforward, but has the disadvantage of damaging both the sample and the tip, due to the strong lateral forces. Therefore another mode has been invented—tapping mode, in which during the lateral scan, the cantilever is oscillated at its resonance frequency in vertical direction. This reduces the damage on the sample and the tip and allows lock-in detection of changes in oscillation amplitude, frequency, phase, thus providing more accurate control of the sample-probe separation. These feedback methods give better spatial resolution and stability than the one achievable with shear force feedback.

A significant drawback of both types of aperture probes is that an unpredictable shift between the topography and the SNOM image in the order of 50–100 nm can easily appear. It originates from the fact that the optical image is obtained through the center of the aperture, while the topography signal is acquired with the most protruding part of the tip, which can be anywhere, especially if a random particle is picked up by the tip during the scan.

Aperture SNOM also has its resolution limitations—since the resolution depends on the hole size, the aperture should be made as small as possible. However, the transmitted light intensity diminishes as a^6 when squeezing the diameter, a, of the hole. Therefore, there is always a compromise between the signal intensity and the lateral resolution. The best lateral resolution achieved with aperture SNOM with still reasonable signal intensity is on the order of 50 nm.

Apertureless Probes

A scattering type of SNOM (apertureless SNOM) has been introduced to overcome the above mentioned resolution limitations of the aperture SNOM. It uses sharp silicon [66] or metal [67] tips to scatter the electromagnetic field on the sample surface during the scan. Since these probes are much sharper than the other two types, much better spatial resolution can be achieved both in the topography and the SNOM image—on the order of 10 nm [45, 46].

Also, both the optical and the topography images originate from the same point, therefore shift between them is unlikely to occur. However the useful near-field signal should be extracted from a huge far-field background and topography induced artifacts might be expected. This requires more complicated detection schemes (for example lock-in) and careful data analysis [44, 68–70].

Other SNOM Probes

All mentioned probes have their advantages and disadvantages and the choice of probe depends on the specific information to be obtained. Numerous dedicated probes have been developed for specific needs. A few interesting examples are briefly mentioned below.

SNOM tips with different active nano-optical structures attached to them have been developed—for example single fluorescent molecule [71], fluorescent quantum dots [72], photodiode [73]. The signal from the light emitting structures is used to map the local electromagnetic field near the sample surface.

Frey et al. have introduced a tip on aperture (TOA) design, by attaching a metallic antenna to a standard optical fiber probe—Fig. 1.12a [74]. The light which is confined by the fiber is further concentrated by the antenna. Thus, such a probe has the resolution of a scattering type SNOM, but introduces less far-field background in the signal. The authors have demonstrated resolution of 25 nm when imaging fluorescent beads.

Special probes have been developed for studying dynamical aspects of light emission and non linear optical processes. An SEM view of such a probe consisting of tapered optical fiber, carbon nanotube (CNT) and functional nano-optical structures is shown in Fig. 1.12b [75].

If a sharp tip is metalized, lightening rod effect or excitation of plasmonic resonances might occur at the tip end. Both lead to enhancement of the electromagnetic field near the tip. These enhanced fields have been demonstrated to excite multiphoton luminescence [76] and second harmonic generation [77]. Yang et al. have demonstrated apertureless SNOM based on the effect of fluorescent quenching near a metalized AFM tip [77]. All these near field probes are explored in the search for better resolution and less far-field background in the image.

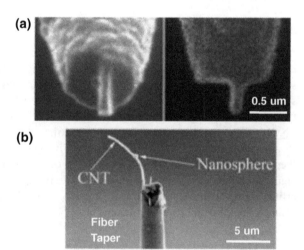

Fig. 1.12 Special types of probes have been developed for enchancing the SNOM resolution and reducing the far-field background contribution to the optical signal like: **a** Tip on aperture (TOA) design in which a metallic antenna is attached to a standard optical fiber probe [74] and **b** Tapered optical fiber, combined with carbon nanotube (CNT) and functional nano-optical structures [75]

1.2.4 Artifacts and Challenges

The main challenges in the SNOM field remain the interpretation of the images, the extraction of the useful near-field information from the far-field background and the presence of artifacts in the optical images. Although the SNOM involves well-known optical phenomena such as transmission, reflection, absorption, etc., additional effects may appear in the near-field. Care must be taken to avoid or at least recognize artifacts in the near-field images and ensure that the optical image is indeed "real". Hecht et al. have warned that the early work in the field of SNOM was very much dominated by artifacts and false claims for nanometer resolution have been made [78]. Listed below are some of the most common SNOM artifacts:

Topographical Artifacts

In a near-field measurement, optical and topographic information can easily be coupled. Two of the main mechanisms playing a role are due to changes of the optical alignment of the system and variations of the near-field in the vicinity of the probe when topographic features are scanned [79].

The first effect can occur if changes in the optical path of the system excitation source-probe-detector occur. In lateral direction this is unlikely, since usually the sample is scanned and the mentioned parts of the system are fixed. However movement of the probe in vertical direction when encountering a topographic feature may shift the focus of the system. This shift is usually minor, but if a confocal detection scheme is used, it may still significantly influence the optical signal.

The origin of the second mechanism is depicted in Fig. 1.13. The optical signal in the SNOM measurement is very sensitive to the material in the immediate vicinity of the sample. If a step is scanned, the material close to the aperture changes from substrate to air and then to the sample material, instead of directly changing from substrate to sample. This may lead to the optical signal drop depicted in Fig. 1.13.

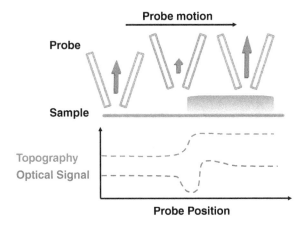

Fig. 1.13 Artifacts in the optical signal due to topography variations may appear in the SNOM measurements

Various mechanisms to recognize topography-induced artifacts have been proposed [78]. For example, a correlation image might be done between the topography and the optical image. If there is a good correlation, then most probably the contrast in the optical image is due to topography. There is however still no clear and well established way to separate and distinguish topographical from optical information.

Probe-Sample Interactions

Another mechanism of artifacts generation is sample perturbations by the tip. As already mentioned, unwanted hard contact between the tip and the sample might result in damaging of both of them. Since the tip is usually metal coated, electrical perturbations of the sample might also occur or electric charges can be transferred between the probe and the sample. For example, it has been shown that excited state life time of a single molecule measured with a tapered optical fiber SNOM probe depends on the position of the molecule relative to the glass and the metal part of the probe [80].

Probe Heating

To achieve better resolution with aperture SNOM, the size of the aperture should be as small as possible. However, this reduces significantly the amount of light passing through the aperture, so the illumination intensity should be increased to attain measurable signal levels. This strong illumination inevitably leads to heating of the probe, mainly due to absorption of light in the metal (typically Al) coating layer and changes in the properties of the probe [81].

For example, the thermal expansion of the probe leads to reduction of the diameter and since the throughput of the optical fiber depends on the sixth power of the aperture's diameter, this effect might significantly lower the throughput. For optical fiber probes, if elongation of the probe occurs, the waveguide modes inside the fiber are attenuated which results in lower throughput as well. With increasing the temperature, the reflectivity of Al decreases, which opens a positive feedback for further heating of the probe. If high temperatures are reached, not only the probe but also the sample could be damaged. The exact damaging conditions depend very much on the concrete experimental configuration and SNOM probe.

1.3 SNOM for Imaging the Magnetic Field of Light

In the first section we have discussed the promising applications offered by the field of plasmonics and that for the further development of plasmonic devices it is crucial to characterize the light-matter interaction in the near-field of the sample with nanoscale resolution. It has been pointed out that one of the critical steps in

1.3 SNOM for Imaging the Magnetic Field of Light

the characterization of plasmonic devices is imaging of the different electric and magnetic field components of the light with nanoscale resolution.

In the second section, we have explained why the characterization of plasmon effects is challenging from experimental point of view and we have discussed that one of the most suitable techniques for studying them is the SNOM. Nowadays, one of the last standing gaps in the full electromagnetic field characterization is the imaging of the magnetic field component of light. The work presented in this thesis contributes to fill in this gap.

In the first part of this section we explain how the various SNOM configurations and probes can image some of those electromagnetic field components. In the second part of the section we will elaborate on the fundamental difficulties in the magnetic field imaging at optical frequencies. Finally, we will describe the experimental setup, which allows the imaging of the magnetic field of light.

1.3.1 Imaging the Electromagnetic Field Components by SNOM

Different combinations of the SNOM probes and configurations described in the previous section result in imaging of the different electromagnetic field components of the light. For example, the normal to the sample surface electric field component can be mapped by an apertureless SNOM configuration [42, 70]. Images obtained with a metal coated optical fiber probe in transmission configuration have been interpreted in terms of local density of optical states (LDOS) [82–84], magnetic field [29, 30, 32, 58, 85] or lateral electric field [86, 87]. However, no consensus has been reached and the imaged component strongly depends on tiny details in the probe geometry.

While imaging of the electric field components is nowadays a standard procedure, mapping of the magnetic field components remains a challenging non trivial task. This is mainly because of the weak interaction of the magnetic field of light with conventional materials, the physical reason for which will be elaborated in the next section [26, 27]. Although indirect procedures, relying on the calculation of the magnetic field from the electric field via Maxwell's equations [31, 88–90] have been suggested, direct measurements of the magnetic field remain difficult. Important progress in this direction has been reported by Burresi et al. [28] who have developed a special split-ring-like probe for mapping the *vertical* magnetic field component. Instead, in our work we focus on imaging the *lateral* magnetic field component of the light, which will be discussed in the last section of this chapter.

1.3.2 Fundamental Challenges in Imaging the Magnetic Field of Light

Since the proposition by James Clerk Maxwell and the experimental confirmation by Hertz, we know that light is an electromagnetic wave—it consists of oscillating electric and magnetic fields, related to each other via Maxwell's equations. In an electromagnetic wave, travelling through a media with refractive index n, the ratio between the electric and magnetic field intensity is:

$$\frac{E}{B} = \frac{c}{n} \tag{1.8}$$

When a material is illuminated with light, those electric and magnetic fields interact with the material to yield a certain response. The fields can interact in various ways with the atoms, electrons, nuclei. Below we will make a rough estimate on the interaction strength of the electric and magnetic field forces, which those particles exhibit.

An electric field with magnitude E acts upon a charged particle with charge q with a force:

$$F_e = qE \tag{1.9}$$

A magnetic field with magnitude B acts upon a moving charge q at speed v with a Lorentz force:

$$F_b = qvB \tag{1.10}$$

Thus, the ratio between the electric and the magnetic force strength is:

$$\frac{F_e}{F_b} = \frac{E}{vB} = \frac{c}{v} \tag{1.11}$$

Since for most materials the refractive index n is on the order of 1 to 2, it has been omitted in the formula above, which has the purpose to provide a rough estimation of the order of magnitude.

If we are considering a metallic material, the free electrons in it are moving roughly at the Fermi velocity [26]. For typical metals it is in the order of 10^6 m/s, consequently:

$$\frac{c}{v} = 300 \tag{1.12}$$

Thus, for the free electrons in a metal, the ratio between the electric and the magnetic force strength is in the order of:

$$\frac{F_e}{F_b} = \frac{c}{v} \approx 300 \tag{1.13}$$

1.3 SNOM for Imaging the Magnetic Field of Light

In addition electrons localized at the atoms' orbitals can also have a certain contribution to the overall response of a material. An estimation about the strength of this interaction can be derived from the Hydrogen atom model for an orbital with radius r_B. According to that model, the kinetic and the potential energies of an electron, orbiting around the nuclei have to be comparable:

$$\frac{mv^2}{2} = \frac{e^2}{4\pi\epsilon_0 r_B} \tag{1.14}$$

The uncertainty principle requires that:

$$mv = \frac{\hbar}{r_B} \tag{1.15}$$

where $\hbar$ is Planck's constant.

From the two equations we can derive an estimate of the ratio of the speed of the electron in the atom and the speed of light (the factor of 1/2 has been neglected):

$$v = \frac{ce^2}{4\pi\epsilon_0 \hbar c} = \alpha c \tag{1.16}$$

The constant alpha is the so-called fine-structure constant:

$$\alpha = \frac{1}{137.036} \tag{1.17}$$

Thus, for the orbiting electrons:

$$\frac{F_e}{F_b} = \frac{c}{v} \approx 137 \tag{1.18}$$

This is a rough estimate for a Hydrogen atom model and the exact numbers vary depending on the specific material. In general, the magnetic force is typically three to four orders of magnitude weaker than the electric one [27].

The magnetic field can also yield certain nuclear magnetic resonances and electron spin resonances. The frequencies of those resonances are however typically in the radio-wave and microwave range, respectively, far below optical frequencies.

This means that for most materials the magnetic light-matter interactions are first, of little importance and second, very difficult to measure. However, as explained in the first chapter, for artificial metamaterials, the magnetic light-matter interactions become as important as the electric ones. Therefore, measuring of the magnetic field of light becomes a necessity for engineering such devices.

The measurements of the magnetic field are based on structuring of the near-field probe, so that it can exhibit a certain magnetic dipole moment. Burresi et al. have shown that a split-ring probe has a magnetic dipole moment normal to the sample surface and it can therefore be used to image the normal component of the magnetic field of light [28].

In our work, we use the fact that a hole in a metal film (Bethe-aperture) can be approximated by a tangential magnetic dipole and respectively, it can be used for detecting the lateral magnetic field of light. The imaging of the magnetic field of light with such aperture probe SNOM will be described in details in Chaps. 2–4 of the thesis. In the last part of the introduction we will give experimental details about the SNOM setup we use.

1.3.3 Experimental Setup for Imaging the Magnetic Field of Light

A description of the SNOM [63] used for imaging the lateral magnetic field of light in this thesis, is presented in this section and illustrated in Fig. 1.14.

A supercontinuum white light (SCWL) laser (NKT Photonics, Koheras SuperK Extreme Standard) with spectral range 400–2000 nm is used as the light source. It delivers up to 5 W of power at 80 MHz repetition rate. From the broad spectrum, single wavelengths with a linewidth of ∼5–20 nm and power of a few mW, depending on the wavelength, are selected with an acousto-optic tunable filter (AOTF) (NKT

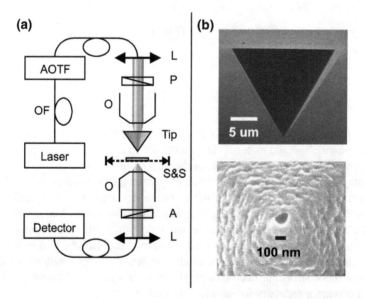

Fig. 1.14 Schematics of the hollow-pyramid SNOM setup used in the thesis for imaging the lateral magnetic field of light. **a** The setup was used in transmission configuration. OF: optical fiber, L: lens, P: polarizer, O: objective, A: analyser, S&S: sample and scan stage. **b** SEM image of the cantilevered aperture probe—side view (*left*) and front view (*right*)

Photonics, SpectraK Dual). Simultaneous selection of up to eight different wavelengths is possible. Since the system is fully fiber coupled, minimal alignment adjustments are required when changing between different AOTF filters and detectors for measurements in different spectral regions. The selected illumination is coupled into the microscope, where it passes through a polarizer and is then focused on the apex of the SNOM probe by a 20× objective with numerical aperture $NA = 0.4$—Fig. 1.14a.

The probe consists of a hollow SiO_2 pyramid, covered with an Al layer with thickness of about 100 nm, which is etched through at the apex. This opens up an aperture with a sub-wavelength diameter of nominally 100 nm, which provides optical resolution below the diffraction limit. Scanning electron micrographs (SEM) of such a probe, showing the pyramidal shape, cantilever, and aperture, are shown in Fig. 1.14b.

Next, the light transmitted through the sample is collected in the far-field by a 60× achromat microscope objective ($NA = 0.8$), directed through an analyzer, and confocally picked up by the pinhole of an optical fiber. The collected light in the fiber can be coupled to different detectors. For the measurements presented in this thesis, two spectrometers, equipped with a front-illuminated Si CCD camera for the visible, and an InGaAs detector array for the near-infrared spectral range, were used. While the sample is scanned, the spectrometer collects a full spectrum at each scan pixel. Near-field images are then constructed by plotting the transmitted light intensity at each pixel for a selected wavelength range only.

Typically areas of 10 μm by 10 μm were measured with a 50 nm pixel size. One scan takes roughly 25 min and typically four to six different excitation wavelengths are measured simultaneously, although even eight is feasible.

The sample can be scanned in atomic force microscopy (AFM) contact or tapping mode with an XYZ piezostage, and beam-deflection laser feedback. A dichroic mirror in this stable feedback system makes a narrow spectral window around 1000 nm experimentally not accessible.

In the following chapter it will be demonstrated that this setup is capable of imaging the lateral magnetic near-field of a plasmonic nanobar sample. In Chap. 3 we will elaborate on the physics behind the imaging of the magnetic field. Chapter 4 shows the application of the technique for studying plasmonic antennas with other geometries.

References

1. S.A. Maier, *Plasmonics: Fundamentals and Applications* (Springer, New York, 2007)
2. L. Novotny, N. van Hulst, Antennas for light. Nat. Photonics **5**, 83–90 (2011)
3. M.E. Stewart, C.R. Anderton, L.B. Thompson, J. Maria, S.K. Gray, J.A. Rogers, R.G. Nuzzo, Nanostructured plasmonic sensors. Chem. Rev. **108**, 494–521 (2008)
4. N. Verellen, P. Van Dorpe, C. Huang, K. Lodewijks, G.A.E. Vandenbosch, L. Lagae, V.V. Moshchalkov, Plasmon line shaping using nanocrosses for high sensitivity localized surface plasmon resonance sensing. Nano Lett. **11**, 391–397 (2011)

5. T. Chung, S.-Y. Lee, E.Y. Song, H. Chun, B. Lee, Plasmonic nanostructures for nano-scale bio-sensing. Sensors **11**, 10907–10929 (2011)
6. D.L. Jeanmaire, R.P. Van Duyne, Surface raman spectroelectrochemistry: part i. heterocyclic, aromatic, and aliphatic amines adsorbed on the anodized silver electrode. J. Electroanal. Chem. **84**, 1 (1977)
7. M.G. Albrecht, J.A. Creighton, Anomalously intense Raman spectra of pyridine at a silver electrode. J. Am. Chem. Soc. **99**, 5215 (1977)
8. M. Stockman, Electromagnetic theory of SERS. Top. Appl. Phys. **103**, 47–65 (2006)
9. E. Le Ru, P. Etchegoin, *Principles of Surface Enhanced Raman Spectroscopy and Related Plasmonic Effects* (Elsevier, 2009)
10. M. Haurylau, G. Chen, H. Chen, J. Zhang, N.A. Nelson, D.H. Albonesi, E.G. Friedman, P.M. Fauchet, On-chip optical interconnect roadmap: challenges and critical directions. IEEE J. Sel. Top. Quant. Electron. **12**, 1699 (2006)
11. E. Ozbay, Plasmonics: merging photonics and electronics at nanoscale dimensions. Science **311**, 189 (2006)
12. W.L. Barnes, A. Dereux, T.W. Ebbesen, Surface plasmon subwavelength optics. Nature **424**, 824–830 (2003)
13. K.F. MacDonald, N.I. Zheludev, Active plasmonics: current status. Laser Photonics Rev. **4**, 562–567 (2010)
14. V.K. Valev, A.V. Silhanek, B. De Clercq, W. Gillijns, Y. Jeyaram, X. Zheng, V. Volskiy, O.A. Aktsipetrov, G.A.E. Vandenbosch, M. Ameloot, V.V. Moshchalkov, T. Verbiest, U-shaped switches for optical information processing at the nanoscale. Small **7**, 2573–2576 (2011)
15. C. Tserkezis, N. Papanikolaou, G. Gantzounis, N. Stefanou, Understanding artificial optical magnetism of periodic metal-dielectric-metal layered structures. Phys. Rev. B **78**, 165114 (2008)
16. J.B. Pendry, A.J. Holden, D.J. Robbins, W.J. Stewart, Magnetism from conductors and enhanced nonlinear phenomena. IEEE Trans. Microwave Theory Tech. **47**, 2075–2084 (1999)
17. V.G. Veselago, The electrodynamics of substances with simultaneously negative values of ϵ and μ. Sov. Phys. Usp. **10**, 509 (1968)
18. C.M. Soukoulis, S. Linden, M. Wegener, Negative refractive index at optical wavelengths. Science **315**, 47–49 (2007)
19. V.M. Shalaev, Optical negative-index metamaterials. Nat. Photonics **1**, 41–48 (2007)
20. D.R. Smith, J.B. Pendry, M.C.K. Wiltshire, Metamaterials and negative refractive index. Science **305**, 788–792 (2004)
21. J.B. Pendry, Negative refraction makes a perfect lens. Phys. Rev. Lett. **85**, 3966 (2000)
22. J.B. Pendry, D. Schurig, D.R. Smith, Controlling electromagnetic fields. Science **312**, 1780–1782 (2006)
23. D. Schurig, J.J. Mock, B.J. Justice, S.A. Cummer, J.B. Pendry, A.F. Starr, D.R. Smith, Metamaterial electromagnetic cloak at microwave frequencies. Science **314**, 977 (2006)
24. U. Leonhardt, Optical conformal mapping. Science **312**, 1777 (2006)
25. N.I. Zheludev, Y.S. Kivshar, From metamaterials to metadevices. Nat. Mater. **11**, 917–924 (2012)
26. M. Pelton, G.W. Bryant, *Introduction to Metal-Nanoparticle Plasmonics* (Wiley, 2013)
27. L.D. Landau, E.M. Lifshitz, *Electrodynamics of Continuous Media* (Pergamon, Oxford, 1960)
28. M. Burresi, D. van Oosten, T. Kampfrath, H. Schoenmaker, R. Heideman, A. Leinse, L. Kuipers, Probing the magnetic field of light at optical frequencies. Science **326**, 550–553 (2009)
29. H.W. Kihm, J. Kim, S. Koo, J. Ahn, K. Ahn, K. Lee, N. Park, D.-S. Kim, Optical magnetic field mapping using a subwavelength aperture. Opt. Express **21**, 5625–5633 (2013)
30. B. le Feber, N. Rotenberg, D.M. Beggs, L. Kuipers, Simultaneous measurement of nanoscale electric and magnetic optical fields. Nat. Photonics **8**, 43–46 (2013)
31. R.L. Olmon, M. Rang, P.M. Krenz, B.A. Lail, L.V. Saraf, G.D. Boreman, M.B. Raschke, Determination of electric-field, magnetic-field, and electric-current distributions of infrared optical antennas: a near-field optical vector network analyzer. Phys. Rev. Lett. **105**, 167403 (2010)

32. D. Denkova, N. Verellen, A.V. Silhanek, V.K. Valev, P. Van Dorpe, V.V. Moshchalkov, Mapping magnetic near-field distributions of plasmonic nanoantennas. ACS Nano **7**, 3168–3176 (2013)
33. E. Abbe, Beiträge zur Theorie des Mikroskops und der mikroskopischen Wahrnehmung. Arch. Mikrosk. Anat. **9**, 413–468 (1873)
34. J. Huisken, D.Y.R. Stainier, Selective plane illumination microscopy techniques in developmental biology. Development **136**, 1963–1975 (2009)
35. D.B. Murphy, M.W. Davidson, *Fundamentals of Light Microscopy and Electronic Imaging* (Wiley-Blackwell, 2012)
36. S.W. Hell, R. Schmidt, A. Egner, Diffraction-unlimited three-dimensional optical nanoscopy with opposing lenses. Nat. Photonics **3**, 381–387 (2009)
37. X. Zhuang, Nano-imaging with STORM. Nat. Photonics **3**, 365–367 (2009)
38. R. Heintzmann, M.G.L. Gustafsson, Subdiffraction resolution in continuous samples. Nat. Photonics **3**, 362–364 (2009)
39. S. Kawata, Y. Inouye, P. Verma, Plasmonics for near-field nano-imaging and superlensing. Nat. Photonics **3**, 288–394 (2009)
40. E.H. Synge, A suggested model for extending microscopic resolution into the ultra-microscopic region. Phil. Mag. **6**, 356–362 (1928)
41. E.A. Ash, G. Nicholls, Super-resolution aperture scanning microscope. Nature **237**, 510–512 (1972)
42. D.W. Pohl, W. Denk, M. Lanz, Optical stethoscopy: image recording with resolution $\lambda/20$. Appl. Phys. Lett. **44**, 651 (1984)
43. B. Hecht, B. Sick, U.P. Wild, V. Deckert, R. Zenobi, O.J.F. Martin, D.W. Pohl, Scanning near-field optical microscopy with aperture probes: fundamentals and applications. J. Chem. Phys. **112**, 7761–7774 (2000)
44. J.-J. Greffet, R. Carminati, Image formation in near-field optics. Prog. Surf. Sci. **56**, 133 (1997)
45. R. Hillenbrand, F. Keilmann, Material-specific mapping of metal/semiconductor/dielectric nanosystems at 10 nm resolution by backscattering near-field optical microscopy. Appl. Phys. Lett. **80**, 25 (2002)
46. A. Bek, R. Vogelgesang, K. Kern, Apertureless scanning near field optical microscope with sub-10 nm resolution. Rev. Sci. Instrum. **77**, 043703 (2006)
47. L. Rayleigh, On the theory of optical images with special reference to the optical microscope. Phil. Mag. **5**, 167–195 (1896)
48. M. Born, E. Wolf, *Principles of Optics*, 7th edn. (Cambridge University Press, Cambridge, 1999)
49. Molecular Expressions (2014), http://micro.magnet.fsu.edu/primer/anatomy/numaperture.html
50. A. Rudge, K. Milne, A. Olver, P. Knight (eds.), *The Handbook of Antenna Design*, vol. 1 (Peter Peregrinus Ltd., London, UK, 1982)
51. G.B. Airy, On the diffraction of an object-glass with circular aperture. Camb. Phil. Soc. Trans. **5**, 283–291 (1835)
52. H. Bethe, Theory of diffraction by small holes. Phys. Rev. **66**, 163–182 (1944)
53. C.J. Bouwkamp, Diffraction theory. Rep. Prog. Phys. **17**, 35 (1954)
54. C. Obermüller, K. Karrai, G. Kolb, G. Abstreiter, Transmitted radiation through a subwavelength-sized tapered optical fiber tip. Ultramicroscopy **61**, 171–177 (1995)
55. A. Lewis, M. Isaacson, A. Harootunian, A. Murray, Development of a 500 Å spatial resolution light microscope. Ultramicroscopy **13**, 227–232 (1984)
56. K. Imura, H. Okamoto, Reciprocity in scanning near-field optical microscopy: illumination and collection modes of transmission measurements. Opt. Lett. **31**, 1474–1476 (2006)
57. E. Méndez, J.-J. Greffet, R. Carminati, On the equivalence between the illumination and collection modes of the scanning near-field optical microscope. Opt. Commun. **142**, 7–13 (1997)
58. D. Denkova, N. Verellen, A.V. Silhanek, P. Van Dorpe, V.V. Moshchalkov, Near-field aperture-probe as a magnetic dipole source and optical magnetic field detector (2014)
59. Cavendish Instruments SNOM Information (2014), http://lamp.tu-graz.ac.at/hadley/nanoscience/week3/3.html

60. D. Richards, Near-field microscopy: throwing light on the nanoworld. Phil. Trans. R. Soc. Lond. A **361**, 2843–2857 (2003)
61. NanoScan Technology (2014), http://www.nanoscantech.ru/en/products/spm/spm-151.html
62. Nanonics SNOM probes (2014), http://www.scientec.fr
63. Witec Wissenschaftliche Instrumente und Technologie GmbH (2014), http://www.witec.de
64. The atomic force microscopy resource library, (2014) http://www.afmuniversity.org
65. M. Celebrano, P. Biagioni, M. Zavelani-Rossi, D. Polli, M. Labardi, M. Allegrini, M. Finazzi, L. Duò, G. Cerullo, Hollow-pyramid based scanning near-field optical microscope coupled to femtosecond pulses: a tool for nonlinear optics at the nanoscale. Rev. Sci. Instrum. **80**, 033704 (2009)
66. F. Zenhausern, M.P. O'Boyle, H.K. Wickramasinghe, Apertureless near-field optical microscope. Appl. Phys. Lett. **65**, 1623 (1994)
67. Y. Inouye, S. Kawata, Near-field scanning optical microscope with a metallic probe tip. Opt. Lett. **19**, 159 (1994)
68. A. García-Etxarri, I. Romero, J.F. García de Abajo, R. Hillenbrand, J. Aizpurua, Influence of the tip in near-field imaging of nanoparticle plasmonic modes: weak and strong coupling regimes. Phys. Rev. B **79**, 125439 (2009)
69. M. Schnell, A. García-Etxarri, J. Alkorta, J. Aizpurua, R. Hillenbrand, Phase-resolved mapping of the near-field vector and polarization state in nanoscale antenna gaps. Nano Lett. **10**, 3524–3528 (2010)
70. L. Novotny, S.J. Stranick, Near-field optical microscopy and spectroscopy with pointed probes. Annu. Rev. Phys. Chem. **57**, 303–331 (2006)
71. J. Michaelis, C. Hettich, J. Mlynek, V. Sandoghdar, Optical microscopy using a single-molecule light source. Nature **405**, 325 (2000)
72. Y. Sonnefraud, N. Chevalier, J.F. Motte, S. Huant, P. Reiss, J. Bleuse, F. Chandezon, M.T. Burnett, W. Ding, S.A. Maier, Near-field optical imaging with a CdSe single nanocrystal-based active tip. Opt. Express **14**, 10596 (2006)
73. S. Akamine, H. Kuwano, H. Yamada, Scanning near-field optical microscope using an atomic force microscope cantilever with integrated photodiode. Appl. Phys. Lett. **68**, 579 (1996)
74. H.G. Frey, F. Keilmann, A. Kriele, R. Guckenberger, Enhancing the resolution of scanning near-field optical microscopy by a metal tip grown on an aperture probe. Appl. Phys. Lett. **81**, 5030 (2002)
75. Y. Jia, H. Li, B. Zhang, X. Wei, Z. Zhang, Z. Liu, Y. Xu, Development of functional nanoprobes for optical near-field characterization. J. Phys. Condens. Matter **22**, 3342218 (2010)
76. E.J. Sánchez, L. Novotny, X.S. Xie, Near-field fluorescence microscopy based on two-photon excitation with metal tips. Phys. Rev. Lett. **82**, 4014 (1999)
77. A.V. Zayats, V. Sandoghdar, Apertureless scanning near-field second-harmonic microscopy. Opt. Commun. **178**, 245 (2000)
78. B. Hecht, H. Bielefeldt, Y. Inouye, D.W. Pohl, L. Novotny, Facts and artifacts in near-field optical microscopy. J. Appl. Phys. **81**, 2492 (1997)
79. D. Bonnell, *Scanning Probe Microscopy and Spectroscopy—Theory, Techniques and Applications* (Wiley, 2001)
80. C. Girard, O. Martin, A. Dereux, Molecular lifetime changes induced by nanometer scale optical fields. Phys. Rev. Lett. **75**, 3098 (1995)
81. A.H. La Rosa, B.I. Yakobson, H.D. Hallen, Origins and effects of thermal processes on near-field optical probes. Appl. Phys. Lett. **67**, 2597 (1995)
82. K. Imura, T. Nagahara, H. Okamoto, Near-field optical imaging of plasmon modes in gold nanorods. J. Chem. Phys. **122**, 154701 (2005)
83. A. Dereux, C. Girard, J.-C. Weeber, Theoretical principles of near-field optical microscopies and spectroscopies. J. Chem. Phys. **112**, 7775 (2000)
84. G. Colas des Francs, C. Girard, J.-C. Weeber, A. Dereux, Relationship between scanning near-field optical images and local density of photonic states. Chem. Phys. Lett. **345**, 512–516 (2001)

References

85. E. Devaux, A. Dereux, E. Bourillot, J.-C. Weeber, Y. Lacroute, J.-P. Goudonnet, C. Girard, Local detection of the optical magnetic field in the near zone of dielectric samples. Phys. Rev. B **62**, 10504 (2000)
86. J.-S. Bouillard, S. Vilain, W. Dickson, A.V. Zayats, Hyperspectral imaging with scanning near-field optical microscopy: applications in plasmonics. Opt. Express **18**, 16513 (2010)
87. S.I. Bozhevolnyi, Near-field mapping of surface polariton fields. J. Microsc. **202**, 313–319 (2001)
88. A. Bitzer, H. Merbold, A. Thoman, T. Feurer, H. Helm, M. Walther, Terahertz near-field imaging of electric and magnetic resonances of a planar metamaterial. Opt. Express **17**, 3826–3834 (2009)
89. T. Grosjean, I.A. Ibrahim, M.A. Suarez, G.W. Burr, M. Mivelle, D. Charraut, Full vectorial imaging of electromagnetic light at subwavelength scale. Opt. Express **18**, 5809–5824 (2010)
90. M. Seo, A.J.L. Adam, J.H. Kang, J. Lee, S.C. Jeoung, Q.H. Park, P.C.M. Planken, D.S. Kim, Fourier-transform terahertz near-field imaging of one-dimensional slit arrays: mapping of electric-field-, magnetic-field-, and Poynting vectors. Opt. Express **15**, 11781 (2007)

Chapter 2
Imaging the Magnetic Near-Field of Plasmon Modes in Bar Antennas

Abstract In this chapter, we show how the scanning near-field optical microscopy (SNOM) technique can be used to visualize the lateral magnetic near-fields of metallic nanostructures, namely gold bars. We present direct experimental maps of these fields by using hollow-pyramid aperture probe SNOM. The results are supported by numerical simulations in which we first simulate the fields of the probe and the bars separately. Then we simulate and discuss in details how the probe-sample interaction results in the effective formation of a lateral magnetic dipole. This allows obtaining optical contrast in the SNOM images corresponding to the lateral magnetic near-fields of the structures. We verify the results for different bar lengths and wavelengths, respectively different plasmon modes. The obtained specific relation of the bar length versus resonant wavelength (so called dispersion relation), allows to unambiguously confirm that the observed optical contrast is related to plasmonic effects.

2.1 Introduction

As discussed in Chap. 1, photonic nanomaterials, and in particular plasmonic nanoantennas, enable light [1–5] and matter [6] manipulation at the nanoscale. Many of the newly emerging fields in photonics are relying on plasmonic devices as essential building blocks, for example—all-optical signal processing devices [7–11], metamaterials [12, 13], ultrahigh sensitivity bio- and chemical- sensors [14–18], and active photodetectors [19, 20], to name a few.

One of the key factors determining the functionality of these photonic devices is the local distribution of the electric and magnetic field components in the vicinity of the nanostructures' boundaries. As mentioned in the introduction, in metamaterials for example, both the electric [21] and the magnetic [22] interactions between the artificial atoms play a crucial role in obtaining negative permittivity and negative

---

The results presented in this chapter are based on and reproduced with permission from:
D. Denkova, N. Verellen, A.V. Silhanek, V.K. Valev, P. Van Dorpe, V.V. Moshchalkov
Mapping magnetic near-field distributions of plasmonic nanoantennas
ACS Nano **7**, 3168 (2013). Copyright © 2013, American Chemical Society.

permeability, which are necessary for the design and the engineering of, e.g., optical cloaking [23] and negative refractive index materials [13, 24]. Recently, a lot of efforts have been concentrated on achieving magnetic field enhancement at optical frequencies [25–27] with prospective applications [28], for example, as magnetic sensors [29, 30] and for achieving magnetic non-linear effects [31]. Therefore, both from fundamental science and applications point of view, mapping not only the electric but also magnetic near-field distributions has become of crucial importance.

Realizing that with standard far-field optical microscopy methods is impossible, as they have insufficient resolution and provide no information about the electromagnetic near-fields [32, 33]. In recent years, several techniques—each with its own specific scope, advantages and restrictions—have been developed and optimized to gain access to the optical near-fields, *e.g.*: cathodoluminescence (CL) [34], electron energy-loss spectroscopy (EELS) [35], two photon luminescence (TPL) [36], second harmonic generation (SHG) microscopy [6], and scanning near-field optical microscopy (SNOM) [29, 37–39].

In this thesis, we focus on the SNOM technique [40, 41]. It relies on scanning different types of probes in the near-field of a sample, which allows imaging different components of the electromagnetic near-field. For example, the vertical electric field component (relative to the sample surface) can be mapped by the sharp needle of a scattering-SNOM [42, 43]. On the other hand, images obtained with a metal coated optical fiber probe, having a sub-wavelength hole at its apex (aperture-SNOM), have been interpreted in terms of local density of optical states (LDOS) [38, 44, 45], magnetic field [30, 46], or lateral electric field [47, 48]. However, no consensus has been reached regarding the image interpretation and the obtained results might be sensitive to small variations in the probe structure and geometry.

While imaging of the electric field components is nowadays considered a standard procedure, mapping of the much more weakly interacting magnetic field components [49] remains a challenging non-trivial task. Indirect procedures have been explored, relying on the calculation of the magnetic field components from the electric field components via Maxwell's equations [50–53]. However, direct measurements of the magnetic field at optical frequencies remain difficult. Significant progress in this direction has been reported by Burresi et al. [29], who have developed a special split-ring like probe for mapping the *vertical magnetic* field component.

Instead, here we demonstrate mapping of the *lateral magnetic* near-field distribution of different plasmon modes in metallic bars, by a hollow-pyramid probe aperture-SNOM (Fig. 2.1a). This type of probes has been used, for instance, for investigating propagating surface plasmon polaritons (SPP) [5, 39, 54, 55]. In these studies, however, the probe-sample coupling and the image contrast formation mechanism, which are crucial for understanding light confining effects in nanoantennas, have not been addressed in details. In this chapter, we propose such a mechanism and illustrate how it effectively results in mapping of the lateral magnetic field distribution of plasmonic nanoantennas.

2.2 Results and Discussion

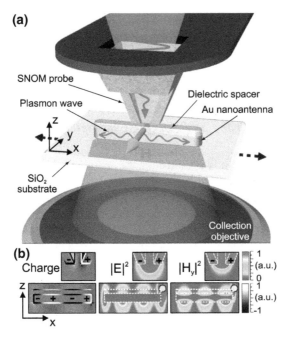

Fig. 2.1 Near-field scanning of a plasmonic nanoantenna with a hollow-pyramid probe. **a** Schematics of the transmission SNOM setup. Electromagnetic coupling between the Al coated hollow-pyramid SNOM probe and the gold nanoantenna induces a lateral magnetic dipole and can excite surface plasmon resonances (SPRs) in the bar. By detecting the transmitted light intensity, while scanning the sample, this dipole allows mapping of SPR magnetic field anti-nodes. **b** Charge density, electric and magnetic field distributions. *Top* at the probe aperture for an x-polarized incident plane wave at $\lambda = 1\,\mu$m. *Bottom* for the $l = 3$ SPR antenna mode ($\lambda = 1270$ nm), excited via an x-polarized dipole source, indicated with the *white circle*. Dimensions of the nanobar are: length $L = 1120$ nm, width $W = 70$ nm and height $h = 50$ nm. The thickness of the dielectric spacer is 30 nm. x-z cross-sections are taken through the probe/sample center

2.2 Results and Discussion

Near-field measurements are performed with a commercial SNOM system—WITec, alpha300 S [56]. The concept of the experiment is illustrated in Fig. 2.1a and a detailed description of the setup is given in Sect. 1.3.3. In short, polarized monochromatic light is focused on the apex of a SNOM probe. This probe consists of a hollow SiO_2 pyramid, mounted on an atomic force microscopy (AFM) cantilever. The pyramid is coated with a 100 nm thick Al layer, which is etched through at the apex. This opens up an aperture with a sub-wavelength diameter of nominally 100 nm, which provides optical resolution below the diffraction limit. For clarity, in Fig. 2.1a, the pyramid is partially cut open.

A part of the incident light can tunnel through the aperture and interact with the sample. The transmitted light is then collected and sent to the detector. Excitation and detection axes are kept collinear while the sample is being scanned in AFM contact mode underneath the probe.

2.2.1 Individual Probe and Sample Characterization

The detected image contrast is a result from the coupling between the near-fields of the probe and the respective near-fields of the sample. Therefore, we first performed a separate study of the probe and the sample. Finite-difference time-domain (FDTD) simulations of the probe reveal that an incident polarized plane wave induces a dipolar charge polarization at its apex—indicated with "+" and "−" signs in Fig. 2.1b, top. This leads to concentration of the electric and magnetic fields, as illustrated for $|E|^2$ and $|H_y|^2$. Due to symmetry considerations, the latter is the only non-zero magnetic field component in the provided x-z cross-section, through the middle of the probe.

All field components, as well as the charge and current density, near the probe aperture are given in the Methods Sect. 2.4, Fig. 2.8. The profiles are obtained at $\lambda = 1$ μm and remain almost unchanged throughout the experimental spectral range. Details on the performed simulations can be found in the Methods Sect. 2.4.

The nanoantenna structure consists of a 50 nm thick and 70 nm wide gold nanobar, covered by a 30 nm thick dielectric layer (see Methods Sect. 2.4, Fig. 2.7). This layer enables scanning of the sample in contact mode while preventing strong conductive coupling between the probe and the sample.

When such a nanobar is illuminated with light, polarized along its long axis, charge density waves at the surface of the metal are excited. They can form standing wave-like Fabry-Pérot resonances, known as surface plasmon resonances (SPRs) [57]. Here, the resonance mode index l is defined as the number of half plasmon wavelengths $\lambda_p/2$ that fit the antenna cavity at resonance.

At positions with high charge density in the bar, strong enhancement of the electric near-field occurs and complementary magnetic maxima appear. This is illustrated in the simulation shown in Fig. 2.1b (bottom) for the $l = 3$ SPR mode, where the mode index is identified from the presented near-field profiles. Here, the antenna is excited by an x-polarized electric dipole, indicated with the white circle. Again, it is worth noticing that H_x and H_z are zero in the central x-z cross-section, as a result of the system's symmetry. Furthermore, the excitation of an SPR leads to resonant enhancement of the antenna's absorption and scattering cross-sections.

2.2 Results and Discussion

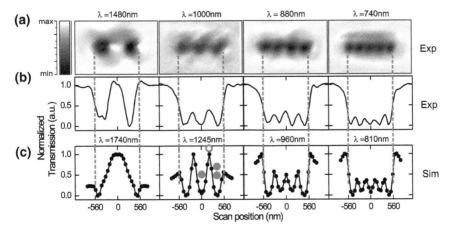

Fig. 2.2 Experimental and simulated near-field transmission scans. **a** Experimental SNOM maps of a nanobar antenna with $L = 1120$ nm at different wavelengths. *Dark* regions indicate reduced transmission intensity. **b** Normalized line scans of the maps in panel a through the center of the antenna. **c** Simulations reproducing the experimental data from panel b. *Red dashed lines* indicate the antenna borders. *Red dots* and *circles* refer to the probe positions, discussed in Fig. 2.3

2.2.2 Probe-Sample Coupling: Imaging of the $|H_y|^2$ Near-Field Distribution of an $l = 3$ Plasmon Mode in a Gold Bar

Now that the sample and probe have been characterized individually, the near-fields induced at the apex of the probe can be used to locally excite SPRs in the near-field region of the nanoantenna.

Experimental near-field scans of a single antenna with length $L = 1120$ nm measured at different excitation wavelengths are shown in Fig. 2.2a. Normalized line scans from the maps in panel a, taken through the center of the bar, are shown in panel b. For each mode, the baseline was subtracted in order to set the transmission minimum to zero, after which the data was normalized to the transmission intensity at the substrate to account for the wavelength dependence of the probe's transmittance [58]. This normalization is chosen to optimize the image contrast, at the expense of losing the absolute transmission intensity information. The light transmitted by the probe is not fully suppressed (*i.e.* not reaching zero) at the transmission minima. Since the nanoantenna is optically thin and narrower than the probe aperture, in addition to the finite extinction cross-section of the SPR resonances, a certain amount of light still reaches the detector. The outer boundaries of the bar are indicated with dashed lines.

Well defined dark spots corresponding to suppressed transmission are observed, while the number of dark spots increases towards shorter wavelengths of the excitation light. From left to right, the strongest transmission contrast was obtained at $\lambda = 1480, 1000, 880$ and 740 nm, respectively. For an accurate determination of these

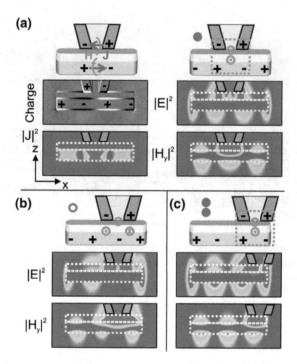

Fig. 2.3 Probe-antenna coupling leads to efficient excitation of SPR only at the lateral magnetic field maxima. **a** *Top* Illustration of image charge formation, induced current J and magnetic field H. *Bottom* simulated charge, current density, $|E|^2$ and $|H_y|^2$ profiles in x-z plane through the middle of the probe and antenna at $\lambda = 1245$ nm with probe at the center of the bar. **b** Simulated $|E|^2$ and $|H_y|^2$ profiles under the same conditions as (**a**), but with probe positioned at 200 nm from the center and **c** near the edge of the antenna. "+" and "−" indicate positive and negative charge accumulation, respectively. *Green circles* indicate out-of-plane magnetic field concentration. In the different panels $|E|^2$ and $|H_y|^2$ profiles have the same color scale, respectively. *Red dots* and *circles* are indicating the position of the tip, shown in Fig. 2.2c. For clarity, the schematic images are not to scale

resonant wavelengths, near-field maps were taken in steps of $\Delta\lambda = 20$ nm and compared. The characteristic features of the experimental near-field images are similar to those of the simulated SPR near-field profiles in Fig. 2.1b and to previous reports with scattering- and optical fiber aperture-SNOM on nanorods [37, 38, 59].

The exact relation between the transmission contrast and SPR near-field distribution is, however, not straightforward and one has to be careful with the interpretation of the SNOM images. We therefore performed extended simulations, including the probe-sample interaction, to reproduce the experimental maps and elucidate the contrast formation mechanisms.

The simulated transmission scans corresponding to Fig. 2.2b taken at, from left to right, $\lambda = 1740, 1245, 960$, and 810 nm, are shown in panel c. These curves were obtained by scanning the probe in 50 nm steps across the antenna while collecting the transmitted light intensity through an area spanning the same angle as the numerical

2.2 Results and Discussion

aperture (NA) of the objective in the measurement. The simulated scans accurately reproduce the experimental number of transmission minima and their relative positions.

The spectral shifts observed between the experimental (panels a and b) and simulated (panel c) transmission resonances most likely originate from deviations from the ideal sample and probe geometry (dimensions, shape, surface roughness, grain boundaries), as well as their optical material properties (depending on fabrication process), used in the simulations. As these parameters are hard to control and accurately determine experimentally, it is difficult to avoid such a spectral mismatch.

We can now confidently use the full-field three dimensional simulations to investigate how the probe fields excite surface plasmons in the nanoantenna, and how this excitation influences the detected transmission intensity. Figure 2.3 provides a more in-depth analysis for the case where three transmission minima are observed in Fig. 2.2c ($\lambda = 1245$ nm) and the probe is located at the same positions as those indicated with red dots and the open circle. In Fig. 2.3a, the probe is positioned above the antenna center and the top left panel illustrates how the polarized charges at the rim of the probe generate image charges in the metallic nanostructure.

This effect is evidenced by the strong electric field intensity $|E|^2$ observed between the Al coating of the probe (grey areas) and the nanobar (also seen in Fig. 2.3b, c). This anti-parallel dipole coupling effectively generates an out-of-plane y-oriented magnetic dipole (green arrows in Figs. 2.1a and 2.3) [60, 61]. The strong lateral magnetic field enhancement is clearly observed in the corresponding $|H_y|^2$ field profiles.

The charge separation induced by the probe, disturbs the free electron gas in the antenna and launches a surface plasmon wave. Even when driven at one of the SPR frequencies, the probe couples to an SPR only at positions for which the waves, reflected back from the antenna edges, are phase matched to form a standing wave pattern. These positions are exactly the nodes in the corresponding SPR charge density distribution.

The charge distribution in Fig. 2.3a clearly shows the standing wave pattern expected for the $l = 3$ antenna mode, exhibiting three charge nodes and *four anti-nodes in the electric field profile*. This resonant charge oscillation translates into three regions of high current density $|J|$ and consequently, through Ampère's law, *three magnetic field anti-nodes*. The top right panel of Fig. 2.3a illustrates how the charges and magnetic near-field (green circles) at the probe aperture line up with the charge distribution and magnetic near-field of the SPR mode.

When resonantly excited, both the absorption and scattering cross-sections of the nanoantenna are drastically enhanced. This means that part of the power transmitted by the probe is absorbed and redirected by the antenna, resulting in a lower detected signal, and therefore, a dark spot in the SNOM image (Fig. 2.2c, single red dot).

When the probe is gradually moved off-center, the symmetric standing wave profile, clearly seen in the $|E|^2$ and $|H_y|^2$ profiles at the bottom surface of the nanobar, is disturbed. Up to a point where one of the four electric field maxima, characteristic for the $l = 3$ mode, disappears. This situation is shown in Fig. 2.3b where the probe is shifted 200 nm from the bar's center. Here, no efficient coupling between the different electric and magnetic field components occurs and excitation of the $l = 3$ mode is not expected. The open red circle in Fig. 2.2c indicates that this results in a transmission maximum.

When approaching the antenna edge (Fig. 2.3c), however, the next $l = 3$ SPR magnetic anti-node is probed. The symmetric SPR field profiles are restored and another transmission minimum is detected (Fig. 2.2c, double red dot).

The closer the probe is to the maximum of the antenna's magnetic field, the stronger the resulting out-of-plane coupled magnetic dipole becomes, leading to stronger excitation of the SPR mode. Consequently, more light is absorbed and less light reaches the detector. From this analysis we can conclude that the SPR field component, effectively mapped in this type of near-field measurement, is H_y.

To further illustrate that the experimental maps indeed correspond to the $|H_y|^2$ field profiles, a comparison between the measured SNOM map and the different components of the electromagnetic near-field of the nanobar is shown in Fig. 2.4. For the simulation, the surface plasmon resonance in the antenna is excited with a dipole source, positioned above one end of the antenna. The profiles are taken at the resonant wavelength for the $l = 3$ SPR mode ($\lambda = 1270$ nm). The electric field components have a common color scale. The magnetic field components also have a common scale bar, except $|H_x|^2$ for which the maximum intensity is an order of magnitude lower.

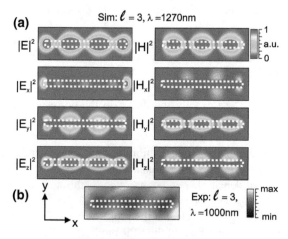

Fig. 2.4 a Electric (*left*) and magnetic (*right*) field distributions for a gold nanobar with $L = 1120$ nm for the $l = 3$ SPR mode ($\lambda = 1270$ nm). x-y cross-sections are taken at 30 nm from the gold top surface. **b** Experimental transmission map for the same mode reproduces the $|H_y|^2$ distribution. The *white boxes* outline the nanobar

2.2.3 Imaging of the $|H_y|^2$ Near-Field Distribution of Different Plasmon Modes in a Gold Bar

Since a transmission minimum was shown to correspond to a region with enhanced magnetic field, it is now possible to unambiguously assign the proper mode index l to the different SPR modes seen in the SNOM maps of Fig. 2.2.

Clearly, the experimental near-field maps reproduce the calculated magnetic field maxima for the $l = 2, 3, 4$, and 5 antenna modes, as illustrated in Fig. 2.5. It is interesting to note that even order modes are also detected [14]. The localized excitation of SPRs offers the necessary symmetry reduction to excite these otherwise dark modes. The simulated profiles here are also obtained by placing a dipole source above one end of the antenna to introduce the phase retardation, which however, leads to a small asymmetry seen in the profiles. From a comparison of the simulation with (Fig. 2.2c) and without (Fig. 2.5 bottom row) the SNOM probe, it can be concluded that the presence of the probe in the near-field of the sample leads to a slight blue-shift of a few percent of the resonant wavelengths. Spectral shifts on the same order of magnitude were observed in other SNOM studies using different types of metal coated and non-metallic probes [62, 63].

2.2.4 Plasmon Dispersion Relation Obtained by SNOM

A typical near-field scan of a nanoantenna array of increasing length L is shown in Fig. 2.6a. Following the dashed arrows, L varies from $L = 720$ nm in the top left to $L = 1800$ nm in the bottom right corner. The width and the height of the bars are kept constant—respectively $W = 70$ nm and $h = 50$ nm. The scan was performed in contact mode at a wavelength of 1100 nm. As L increases, additional half SPP wavelengths can fit the antenna cavity, higher order SPR modes are excited and the number of transmission minima increases (Fig. 2.6).

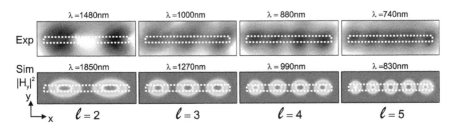

Fig. 2.5 Experimental SNOM transmission maps reproduce the simulated SPR lateral magnetic near-field distributions. Experimental transmission minima seen in Fig. 2.2a, zoomed in in the *top row* here, correspond to the simulated *magnetic near-field maxima* of the $l = 2, 3, 4$, and 5 SPR modes in the nanobar antenna ($L = 1120$ nm), shown in the *bottom row*. Field profiles are taken at 30 nm from the antenna surface. The *white boxes* outline the nanobar

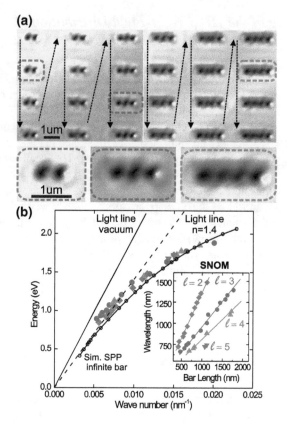

Fig. 2.6 Dispersion relation confirms that SNOM results are related to plasmonic effects. **a** Experimental SNOM scans at $\lambda = 1100$ nm and polarization along the bars' axis of two arrays with bars of increasing length (indicated by the *dashed arrows*). Zoomed in bars with resonant transmission contrast are shown in the *bottom panels*. $L = 750$, 1150 and 1700 nm from *left* to *right*, respectively. **b** Dispersion curves showing resonant plasmon energy versus plasmon wave number k defined as $k = \pi l/L$. Closed symbols: experimental near-field transmission resonances for $l = 2, 3, 4$, and 5 modes. *Solid* and *dashed lines* show the *light line* in vacuum and in n = 1.4 medium, resp. Black circles: calculated plasmon dispersion of an antenna of infinite length L. Inset: the dispersion data displayed as $\lambda(L)$ illustrates the expected linear wavelength scaling behavior for metallic nanorod antennas

Since the SPR modes are spectrally relatively broad, several antennas show the same number of dark transmission spots at a fixed wavelength. To define the resonant antenna length, for a specific SPR mode, again the bar showing the strongest transmission contrast for a line scan through the middle of the bar, was chosen. The resonant antennas in Fig. 2.6a are indicated with dashed boxes and have a length of $L = 750$, 1150 and 1700 nm from left to right, respectively. Note that the systematically observed gradual transition between the plasmon modes excludes possible sample imperfections as the origin of the image contrast and thus, such imperfections, if present, do not affect the presented data analysis and conclusions.

2.2 Results and Discussion

Performing this measurement at different wavelengths allows to map out the SPR dispersion curve, as shown in Fig. 2.6b (closed symbols). The SPR wave number $k = 2\pi/\lambda_p$, with λ_p the SPR wavelength, is defined as $k = \pi l/L$, resulting from the geometrical condition of a standing wave in the antenna cavity [38]. Here, L is the antenna length and l the mode index which was demonstrated to correspond to the number of transmission dips. In defining k a possible systematic error resulting from the non-trivial phase shift of the plasmon wave upon reflection at the bar edges is ignored. This phase shift is introduced by the complex refractive index of the dispersive metallic medium of the plasmon wave [34, 38]. After measuring all points, the first few points and a few random points were remeasured to exclude possible resonance shifts caused by wearing out of the tip.

The solid and dashed lines in Fig. 2.6b are the light lines in vacuum and in a glass medium, with refractive index $n = 1.4$, respectively. Open circles represent the calculated mode dispersion of propagating SPP in the nanoantenna geometry with infinite length L (see Methods Sect. 2.4).

Both the experimental cavity mode dispersion and calculated SPP dispersion bend to the right of the light line at larger wave numbers illustrating the sub-wavelength nature of surface plasmons. This excellent agreement between the dispersion curves further demonstrates that the experimentally observed near-field transmission contrast is indeed mapping plasmonic modes in the nanobar antennas and so justifies the standing wave description [64].

The inset in Fig. 2.6b shows the dispersion data as wavelength *versus* bar length. This graph further illustrates the expected linear wavelength scaling behaviour for metallic nanorod antennas [65–67].

Finally, the presented results can be situated among other reported techniques for obtaining the magnetic field of light. As already pointed out in the introduction, several methods to calculate the magnetic field from the electric field via Maxwell's equations have been developed in the terahertz [52, 53] and optical frequency regions [50, 51]. Although this approach has the advantage of providing information on both the electric and magnetic field components simultaneously (including their phase) it remains as an indirect method which requires post-processing of the data.

One of the most direct ways to measure the optical magnetic field is based on a split ring probe [29]. This is also a phase sensitive method with, in principle, no restrictions concerning the type of optical waves under investigation. However, it has the disadvantage that such probes are not commercially available and they can only access the vertical component of the magnetic field.

Devaux et al. [46] have reported that excitation of resonant circular plasmons in metalized aperture probes can lead to imaging of the magnetic field intensity in dielectric samples. The method is based on resonant effects in the probe and is therefore strongly restricted in terms of illumination wavelength. Another method, allowing simultaneous imaging of the vertical magnetic and electric field components in photonic crystal cavities has recently been reported [68, 69]. It is based on a particular blue-shift induced in the resonant frequencies of those samples and its applicability to other photonic (including plasmonic) systems has not been demonstrated so far.

Most notably, in contrast to the methods mentioned above, the technique we propose here provides information about the *lateral* magnetic field distribution in plasmonic antennas. Our measurements use commercially available probes and do not involve any additional data processing. Moreover, we expect our results to be directly applicable to the more widely used metal coated optical fiber probes, as their geometry is very similar to the one of the hollow-pyramid probes, used in this study. Indications that the metal coated optical fiber probes might indeed be sensitive to the magnetic near-field have been reported [30, 46]. Compared to the optical fiber, the hollow-pyramid probes are very robust [70] (we have experienced that a probe can easily provide good quality images for more than 2–3 months). The probes allow measurements in a broad wavelength region, in our case 500–1600 nm, the limitations coming from the excitation sources and the detectors. As of today, our microscope is not equipped to perform phase-sensitive measurements, but this could be implemented.

It should be emphasized that the presented in this chapter results only demonstrate the applicability of the technique to plasmonic waves in metallic nanobar antennas. Nevertheless, imaging of the lateral magnetic field, in this work, is possible due to the anti-parallel dipole coupling between the separated charges in the probe and the sample, which leads to the effective formation of a lateral magnetic dipole. Therefore, mapping of the lateral magnetic field distribution should be possible in any sample in which such charge separation can be induced, irrespective of the specific sample geometry. We are also optimistic to envisage, that this approach could be extended to propagating surface plasmon polariton waves as well [39]. From this point of view, we believe that our method can be considered as a complementary one to those already reported in the literature.

2.3 Conclusions

In conclusion, we have shown that the lateral magnetic near-field distribution of surface plasmon resonance modes in optical nanobar antennas can be visualized by means of aperture scanning near-field optical microscopy. The formation of an effective magnetic dipole between the hollow-pyramid probe and the antenna was shown to excite standing wave-like surface plasmons in the antenna, only at the SPR lateral magnetic field maxima. This excitation results in a measurable modulation of the transmitted light intensity. These findings are of paramount importance for achieving a complete characterization, including the magnetic field components, of electromagnetic near-field light phenomena mediated by nanoplasmonic devices.

Our findings suggest that aperture-SNOM can be considered as an important complement to the available scattering-SNOM techniques. It would, for example, be the method of choice for near-field studies of optical magnetic field enhancing and confining nanoantennas [25–27].

2.4 Methods

2.4.1 Sample Fabrication

The sample consists of a 150 µm thick glass slide coated with 10 nm of indium tin oxide (ITO) and a thin Ti adhesion layer (Fig. 2.7). A 50 nm thick gold film was sputtered and covered with another Ti adhesion layer and a negative tone hydrogen silsesquioxane (HSQ) resist. The nanostructures were further structured using electron beam lithography and Xe ion milling. The resulting bar width W is ~70 nm.

A residual resist layer with an estimated thickness of 30 nm remains on top of the gold particles and is not removed. Optically, this layer behaves as silicon dioxide and causes the SPR modes to redshift (*e.g.*: 20 nm for the $l = 3$ mode of an $L = 1120$ nm antenna) due to the increase in surrounding refractive index. Additionally, the thickness of the layer will affect the spatial resolution [71] and the probe-sample coupling strength [63]. Although a detailed study of these effects for the present system was not performed, the results are expected to be qualitatively valid for a broad range of layer thicknesses. Most importantly, the layer serves as a dielectric insulator, preventing electrical contact, and therefore conductive coupling, between the nanoparticles and the metallic SNOM probe. This allows us to do fast contact mode scanning of the sample without drastically altering the plasmonic properties. The sample is organized in arrays, which consist of bars with increasing size in steps of 20, 30 or 50 nm.

2.4.2 FDTD Simulations

Simulations were performed with the commercial FDTD Maxwell equations solver Lumerical © [72]. For the near-field simulations, the nanostructure (including the resist layer on top of the metal), the hollow pyramid, and the substrate were placed in an 7 µm × 7 µm × 3.5 µm volume with perfectly matched layer (PML) boundaries and a mesh of 5 nm × 5 nm × 2.5 nm covering the nanostructure and the pyramid's aperture. For excitation, a plane wave source is positioned inside the pyramid.

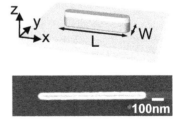

Fig. 2.7 Sketch of the sample structure and SEM image of a single nanobar of length $L = 1120$ nm and width $W = 70$ nm

To reproduce the experimental conditions as accurate as possible, the NA of the collection objective was taken into account by calculating the transmission intensity through a rectangular surface spanning the same collection angle. The permittivity of Au and Al was taken from Refs. [73, 74], respectively. The refractive index of the substrate and resist layer was set to $n = 1.4$. The ITO and Ti layers have a negligible effect on the results and this effect is not included in the simulations.

Field profiles in the absence of the probe were obtained using a point dipole source positioned above the edge of the bar in order to excite both odd and even order modes through phase retardation effects [57]. For the calculation of the plasmon dispersion in Fig. 2.6b the mode solver of Lumerical FDTD was used.

2.4.3 Electric and Magnetic Field Profiles at the Apex of the Probe

The structure of the probe was modelled following the information provided by the manufacturer (materials used, layer thicknesses, cone angle, and aperture size). Schematics of the simulated probe together with the dimensions is shown in Fig. 2.8a. Surface roughness, possible layer thickness variations, and rounding of the edges were not taken into account.

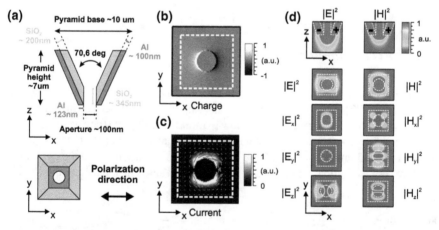

Fig. 2.8 Probe characterization. **a** Schematics of the simulated probe—*side view* (*top*) and *bottom view* (*bottom*). **b** Charge distribution inside the probe 30 nm above the aperture shows dipolar charge separation. **c** Current distribution inside the probe 30 nm above the probe aperture shows that the currents flow in two curved halfs of a *circle*. **d** Electric (*left*) and magnetic (*right*) field distributions at the probe aperture for an x-polarized incident plane wave at $\lambda = 1$ μm. x-z cross-sections taken through the probe center, x-y cross-sections 30 nm below the aperture. *Dashed white lines* indicate probe's boundaries

2.4 Methods

Finite-difference time-domain (FDTD) simulations of the probe, performed with Lumerical © [72], show that an incident x-polarized plane wave induces a dipolar charge polarization at its apex (Fig. 2.8b). The corresponding current profile (Fig. 2.8c), shows that the current flows following the red arrows. As a result, the H_y field is adding up just beneath the probe as is evident from the field profiles shown in Fig. 2.8d. The H_x and H_z fields produced by these currents are zero at a vertical plane through the middle of the probe, along the direction of incident light polarization. The electric and the magnetic field plots have a common color scale.

The profiles are obtained at $\lambda = 1\,\mu m$ and remain almost unchanged throughout the experimental spectral range. For the field profiles, the x-z cross-sections are taken through the probe center and the x-y cross-sections 30 nm below the aperture. For the charge and current profiles, the x-y cross section is taken 30 nm above the aperture.

References

1. L. Novotny, N. van Hulst, Antennas for light. Nat. Photonics **5**, 83–90 (2011)
2. W.L. Barnes, A. Dereux, T.W. Ebbesen, Surface plasmon subwavelength optics. Nature **424**, 824–830 (2003)
3. V. Giannini, A.I. Fernandez-Domínguez, S.C. Heck, S.A. Maier, Plasmonic nanoantennas: fundamentals and their use in controlling the radiative properties of nanoemitters. Chem. Rev. **111**, 3888–3912 (2011)
4. S. Lal, S. Link, N.J. Halas, Nano-optics from sensing to waveguiding. Nat. Photonics **1**, 641–648 (2007)
5. R. Zia, J.A. Schuller, M.L. Brongersma, Near-field characterization of guided polariton propagation and cutoff in surface plasmon waveguides. Phys. Rev. B **74**, 165415 (2006)
6. V.K. Valev, D. Denkova, X. Zheng, A.I. Kuznetsov, C. Reinhardt, B.N. Chichkov, G. Tsutsumanova, E.J. Osley, V. Petkov, B. De Clercq, A.V. Silhanek, Y. Jeyaram, V. Volskiy, P.A. Warburton, G.A.E. Vandenbosch, S. Russev, O.A. Aktsipetrov, M. Ameloot, V.V. Moshchalkov, T. Verbiest, Plasmon-enhanced sub-wavelength laser ablation: plasmonic nanojets. Adv. Mater. **24**, OP29–OP35 (2012)
7. S.A. Maier, P.G. Kik, H.A. Atwater, S. Meltzer, E. Harel, B.E. Koel, A.A. Requicha, Local detection of eletromagnetic energy transport below the diffraction limit in metal nanoparticle plasmon waveguides. Nat. Mater. **2**, 229–232 (2003)
8. V.K. Valev, A.V. Silhanek, B. De Clercq, W. Gillijns, Y. Jeyaram, X. Zheng, V. Volskiy, O.A. Aktsipetrov, G.A.E. Vandenbosch, M. Ameloot, V.V. Moshchalkov, T. Verbiest, U-shaped switches for optical information processing at the nanoscale. Small **7**, 2573–2576 (2011)
9. D. Dregely, R. Taubert, J. Dorfmüller, R. Vogelgesang, K. Kern, H. Giessen, 3D optical Yagi-Uda nanoantenna array. Nat. Commun. **2**, 267 (2011)
10. K.F. MacDonald, N.I. Zheludev, Active plasmonics: current status. Laser Photonics Rev. **4**, 562–567 (2010)
11. L. Yin, V.K. Vlasko-Vlasov, J. Pearson, J.M. Hiller, J. Hua, U. Welp, D.E. Brown, C.W. Kimball, Subwavelength focusing and guiding of surface plasmons. Nano Lett. **5**, 1399–1402 (2005)
12. I.I. Smolyaninov, Two-dimensional plasmonic metamaterials. Appl. Phys. A **87**, 227–234 (2007)
13. V.M. Shalaev, Optical negative-index metamaterials. Nat. Photonics **1**, 41–48 (2007)
14. N.J. Halas, S. Lal, W.-S. Chang, S. Link, P. Nordlander, Plasmons in strongly coupled metallic nanostructures. Chem. Rev. **111**, 3913–3961 (2011)
15. N. Verellen, P. Van Dorpe, C. Huang, K. Lodewijks, G.A.E. Vandenbosch, L. Lagae, V.V. Moshchalkov, Plasmon line shaping using nanocrosses for high sensitivity localized surface plasmon resonance sensing. Nano Lett. **11**, 391–397 (2011)

16. M.E. Stewart, C.R. Anderton, L.B. Thompson, J. Maria, S.K. Gray, J.A. Rogers, R.G. Nuzzo, Nanostructured plasmonic sensors. Chem. Rev. **108**, 494–521 (2008)
17. A.V. Kabashin, P. Evans, S. Pastkovsky, W. Hendren, G.A. Wurtz, R. Atkinson, R. Pollard, V.A. Podolskiy, A.V. Zayats, Plasmonic nanorod metamaterials for biosensing. Nat. Mater. **8**, 867–871 (2009)
18. T. Chung, S.-Y. Lee, E.Y. Song, H. Chun, B. Lee, Plasmonic nanostructures for nano-scale bio-sensing. Sensors **11**, 10907–10929 (2011)
19. M.W. Knight, H. Sobhani, P. Nordlander, N.J. Halas, Photodetection with active optical antennas. Science **332**, 702 (2011)
20. P. Neutens, P. Van Dorpe, I. De Vlaminck, L. Lagae, G. Borghs, Electrical detection of confined gap plasmons in metal-insulator-metal waveguides. Nat. Photonics **3**, 283–286 (2009)
21. J. Zhou, T. Koschny, C.M. Soukoulis, Magnetic and electric excitations in split ring resonators. Opt. Express **15**, 17881–17890 (2007)
22. C.J. Tang, P. Zhan, Z.S. Cao, J. Pan, Z. Chen, Z.L. Wang, Magnetic field enhancement at optical frequencies through diffraction coupling of magnetic plasmon resonances in metamaterials. Phys. Rev. B **83**, 041402 (2011)
23. J.B. Pendry, D. Schurig, D.R. Smith, Controlling electromagnetic fields. Science **312**, 1780–1782 (2006)
24. C.M. Soukoulis, S. Linden, M. Wegener, Negative refractive index at optical wavelengths. Science **315**, 47–49 (2007)
25. T. Grosjean, M. Mivelle, F.I. Baida, G.W. Burr, U.C. Fischer, Diabolo nanoantenna for enhancing and confining the magnetic optical field. Nano Lett. **11**, 1009–1013 (2011)
26. Z. Gao, L. Shen, E. Li, L. Xu, Z. Wang, Cross-Diabolo nanoantenna for localizing and enhancing magnetic field with arbitrary polarization. J. Lightwave Technol. **30**, 829–833 (2012)
27. S. Koo, M.S. Kumar, J. Shin, D. Kim, N. Park, Extraordinary magnetic field enhancement with metallic nanowire: role of surface impedance in Babinet's principle for sub-skin-depth regime. Phys. Rev. Lett. **103**, 263901 (2009)
28. H. Giessen, R. Vogelgesang, Glimpsing the weak magnetic field of light. Science **326**, 529–530 (2009)
29. M. Burresi, D. van Oosten, T. Kampfrath, H. Schoenmaker, R. Heideman, A. Leinse, L. Kuipers, Probing the magnetic field of light at optical frequencies. Science **326**, 550–553 (2009)
30. H. Kihm, S. Koo, Q. Kim, K. Bao, J. Kihm, W. Bak, S. Eah, C. Lienau, H. Kim, P. Nordlander, N. Halas, N. Park, D.-S. Kim, Bethe-hole polarization analyser for the magnetic vector of light. Nat. Commun. **2**, 451 (2011)
31. M.W. Klein, C. Enkrich, M. Wegener, S. Linden, Second-harmonic generation from magnetic metamaterials. Science **313**, 502–504 (2006)
32. E. Abbe, Beiträge zur Theorie des Mikroskops und der mikroskopischen Wahrnehmung. Arch. Mikrosk. Anat. **9**, 413–468 (1873)
33. L. Rayleigh, On the theory of optical images with special reference to the optical microscope. Phil. Mag. **5**, 167–195 (1896)
34. E.J.R. Vesseur, R. de Waele, M. Kuttge, A. Polman, Direct observation of plasmonic modes in Au nanowires using high-resolution cathodoluminescence spectroscopy. Nano Lett. **7**, 2843–2846 (2007)
35. M. Bosman, V.J. Keast, M. Watanabe, A.I. Maaroof, M.B. Cortie, Mapping surface plasmons at the nanometre scale with an electron beam. Nanotechnology **18**, 165505 (2007)
36. P. Ghenuche, S. Cherukulappurath, T.H. Taminiau, N.F. van Hulst, R. Quidant, Spectroscopic mode mapping of resonant plasmon nanoantennas. Phys. Rev. Lett. **101**, 116805 (2008)
37. J. Dorfmüller, R. Vogelgesang, T.R. Weitz, C. Rockstuhl, C. Etrich, T. Pertsch, F. Lederer, K. Kern, Fabry-Pérot resonances in one-dimensional plasmonic nanostructures. Nano Lett. **9**, 2372–2377 (2009)
38. K. Imura, T. Nagahara, H. Okamoto, Near-field opticalimaging of plasmon modes in gold nanorods. J. Chem. Phys. **122**, 154701 (2005)
39. R. Zia, M.L. Brongersma, Surface plasmon polariton analogue to Young's double-slit experiment. Nat. Nanotechnol. **2**, 426–429 (2007)

40. E.H. Synge, A suggested model for extending microscopic resolution into the ultra-microscopic region. Phil. Mag. **6**, 356–362 (1928)
41. D.W. Pohl, W. Denk, M. Lanz, Optical stethoscopy: image recording with resolution $\lambda/20$. Appl. Phys. Lett. **44**, 651 (1984)
42. L. Novotny, S.J. Stranick, Near-field optical microscopy and spectroscopy with pointed probes. Annu. Rev. Phys. Chem. **57**, 303–331 (2006)
43. R. Esteban, R. Vogelgesang, J. Dorfmüller, A. Dmitriev, C. Rockstuhl, C. Etrich, K. Kern, Direct near-field optical imaging of higher order plasmonic resonances. Nano Lett. **8**, 3155–3159 (2008)
44. A. Dereux, C. Girard, J.-C. Weeber, Theoretical principles of near-field optical microscopies and spectroscopies. J. Chem. Phys. **112**, 7775 (2000)
45. G. Colas des Francs, C. Girard, J.-C. Weeber, A. Dereux, Relationship between scanning near-field optical images and local density of photonic states. Chem. Phys. Lett. **345**, 512–516 (2001)
46. E. Devaux, A. Dereux, E. Bourillot, J.-C. Weeber, Y. Lacroute, J.-P. Goudonnet, C. Girard, Local detection of the optical magnetic field in the near zone of dielectric samples. Phys. Rev. B **62**, 10504 (2000)
47. J.-S. Bouillard, S. Vilain, W. Dickson, A.V. Zayats, Hyperspectral imaging with scanning near-field optical microscopy: applications in plasmonics. Opt. Express **18**, 16513 (2010)
48. S.I. Bozhevolnyi, Near-field mapping of surface polariton fields. J. Microsc. **202**, 313–319 (2001)
49. L.D. Landau, E.M. Lifshitz, *Electrodynamics of Continuous Media* (Pergamon, Oxford, 1960)
50. R.L. Olmon, M. Rang, P.M. Krenz, B.A. Lail, L.V. Saraf, G.D. Boreman, M.B. Raschke, Determination of electric-field, magnetic-field, and electric-current distributions of infrared optical antennas: a near-field optical vector network analyzer. Phys. Rev. Lett. **105**, 167403 (2010)
51. T. Grosjean, I.A. Ibrahim, M.A. Suarez, G.W. Burr, M. Mivelle, D. Charraut, Full vectorial imaging of electromagnetic light at subwavelength scale. Opt. Express **18**, 5809–5824 (2010)
52. M. Seo, A.J.L. Adam, J.H. Kang, J. Lee, S.C. Jeoung, Q.H. Park, P.C.M. Planken, D.S. Kim, Fourier-transform terahertz near-field imaging of one-dimensional slit arrays: mapping of electric-field-, magnetic-field-, and poynting vectors. Opt. Express **15**, 11781 (2007)
53. A. Bitzer, H. Merbold, A. Thoman, T. Feurer, H. Helm, M. Walther, Terahertz near-field imaging of electric and magnetic resonances of a planar metamaterial. Opt. Express **17**, 3826–3834 (2009)
54. E. Verhagen, J.A. Dionne, L.K. Kuipers, H.A. Atwater, A. Polman, Near-field visualization of strongly confined surface plasmon polaritons in metal-insulator-metal waveguides. Nano Lett. **8**, 2925–2929 (2008)
55. M. Celebrano, M. Zavelani-Rossi, P. Biagioni, D. Polli, M. Finazzi,L. Duò, G. Cerullo, M. Labardi, M. Allegrini, J. Grand, P. Royer, and P.M. Adam, Mapping local field distribution at metal nanostructures by near-field second-harmonic generation. In: Proceedings of the SPIE, plasmonics: metallic nanostructures and their optical properties V 6641, 66411E–1 2007
56. Witec Wissenschaftliche Instrumente und Technologie GmbH (2014), http://www.witec.de
57. J. Dorfmüller, R. Vogelgesang, W. Khunsin, C. Rockstuhl, C. Etrich, K. Kern, Plasmonic nanowire antennas: experiment, simulation, and theory. Nano Lett. **10**, 3596–3603 (2010)
58. B. Hecht, B. Sick, U.P. Wild, V. Deckert, R. Zenobi, O.J.F. Martin, D.W. Pohl, Scanning near-field optical microscopy with aperture probes: fundamentals and applications. J. Chem. Phys. **112**, 7761–7774 (2000)
59. K. Imura, T. Nagahara, H. Okamoto, Characteristic near-field spectra of single gold nanoparticles. Chem. Phys. Lett. **400**, 500–505 (2004)
60. Y. Ekinci, A. Christ, M. Agio, O.J.F. Martin, H.H. Solak, J.F. Löffler, Electric and magnetic resonances in arrays of coupled gold nanoparticle in-tandem pairs. Opt. Express **16**, 13287 (2008)
61. A. Dmitriev, T. Pakizeh, M. Käll, D.S. Sutherland, Gold-silica-gold nanosandwiches: tunable bimodal plasmonic resonators. Small **3**, 294–299 (2007)

62. J.A. Porto, P. Johansson, S.P. Apell, T. López-Rios, Resonance shift effects in apertureless scanning near-field optical microscopy. Phys. Rev. B **67**, 085409 (2003)
63. A. García-Etxarri, I. Romero, J.F. García de Abajo, R. Hillenbrand, J. Aizpurua, Influence of the tip in near-field imaging of nanoparticle plasmonic modes: weak and strong coupling regimes. Phys. Rev. B **79**, 125439 (2009)
64. G. Schider, J.R. Krenn, A. Hohenau, H. Ditlbacher, A. Leitner, F.R. Aussenegg, W.L. Schaich, I. Puscasu, B. Monacelli, G. Boreman, Plasmon dispersion relation of Au and Ag nanowires. Phys. Rev. B **68**, 155427 (2003)
65. L. Novotny, Effective wavelength scaling for opticalantennas. Phys. Rev. Lett. **98**, 266802 (2007)
66. F. Neubrech, D. Weber, R. Lovrincic, A. Pucci, M. Lopes, T. Toury, M. Lamy de la Chapelle, Resonances of individual lithographic gold nanowires in the infrared. Appl. Phys. Lett. **93**, 163105 (2008)
67. R.L. Olmon, P.M. Krenz, A.C. Jones, G.D. Boreman, M.B. Raschke, Near-field imaging of optical antenna modes in the mid-infrared. Opt. Express **16**, 20295 (2008)
68. S. Vignolini, F. Intonti, F. Riboli, L. Balet, L.H. Li, M. Francardi, A. Gerardino, A. Fiore, D.S. Wiersma, M. Gurioli, Magnetic imaging in photonic crystal microcavities. Phys. Rev. Lett. **105**, 123902 (2010)
69. M. Burresi, T. Kampfrath, D. van Oosten, J.C. Prangsma, B.S. Song, S. Noda, L. Kuipers, Magnetic light-matter interactions in a photonic crystal nanocavity. Phys. Rev. Lett. **105**, 123901 (2010)
70. M. Celebrano, P. Biagioni, M. Zavelani-Rossi, D. Polli, M. Labardi, M. Allegrini, M. Finazzi, L. Duò, G. Cerullo, Hollow-pyramid based scanning near-field optical microscope coupled to femtosecond pulses: a tool for nonlinear optics at the nanoscale. Rev. Sci. Instrum. **80**, 033704 (2009)
71. X. Heng, X. Cui, D.W. Knapp, J. Wu, Z. Yaqoob, E.J. McDowell, D. Psaltis, C. Yang, Characterization of light collection through a subwavelength aperture from a point source. Opt. Express **14**, 10410–10425 (2006)
72. Lumerical Solutions (2014), http://www.lumerical.com
73. P.B. Johnson, R.W. Christy, Optical constants of thenoble metals. Phys. Rev. B **6**, 4370–4379 (1972)
74. D.R. Lide, *CRC Handbook of Chemistry and Physics*, 3rd edn. (CRC, Boca Raton, 2000)

Chapter 3
A Near-Field Aperture-Probe as an Optical Magnetic Source and Detector

Abstract In the previous chapter we have shown that the lateral magnetic near-field distributions of different resonant modes in plasmonic structures can be mapped with a hollow-pyramid aperture SNOM. We have also discussed how the coupling between the probe and the metal bars leads to the obtained results. This chapter focuses on the underlying mechanism for mapping the lateral magnetic field with this circular aperture type probe. We suggest that such probe can be approximated by a lateral magnetic dipole source, which also allows its use as a detector for the lateral magnetic near-field. The equivalence of the reciprocal configurations when the probe is used as a source (illumination mode) and as a detector (collection mode) is experimentally demonstrated for a plasmonic nanobar sample. Verification for dielectric structures remains to be realized. The simplification of the probe to a simple magnetic dipole is extremely useful from a practical point of view, as it facilitates the simulations and the understanding of the near-field images.

3.1 Introduction

As pointed out in the previous chapters, the ability to measure not only the traditionally measured electric components of the electromagnetic field of light, but also the magnetic ones is of tremendous importance for newly appearing classes of photonic devices, such as metamaterials [1–6]. Thus, both optical magnetic field sources and detectors are now required for detailed analysis and further engineering of such materials.

In this chapter we suggest that the stand-alone circular aperture SNOM probe can be considered as a tangential optical magnetic dipole source. Reciprocally, we suggest that the probe can be considered as a tangential optical magnetic field detector. This allows us to substitute the probe in SNOM simulations by an effective magnetic

---

This chapter is based on the following manuscript:
D. Denkova, N. Verellen, A.V. Silhanek, P. Van Dorpe, V.V. Moshchalkov
Near-field aperture-probe as a magnetic dipole source and optical magnetic field detector
Submitted to arXiv:1406.7827 [Physics. Optics] (2014).

dipole, which significantly decreases the calculation time and memory requirements. Our findings are theoretically justified and validated by numerical simulations of the stand-alone probe and the probe-sample interactions in plasmonic samples. Strong magnetic field contribution to the near-field images obtained on dielectric nanophotonic samples using similar type of probes (matellized aperture optical fiber probes) has recently been reported [7, 8]. For the hollow-pyramid probes we use, the verification on other types of samples, such as the above mentioned dielectric samples, is still to be performed.

In the first section we will discuss the underlying physics justifying the approximation of the hollow-pyramid probe to a tangential magnetic dipole. In the next section, we confirm by simulations that the electromagnetic fields of the probe are indeed resembling the ones of a lateral magnetic dipole. Then, we demonstrate that in a simulation describing the real experiment, in which the probe is scanned over a sample (demonstrated for plasmonic bar), similar results are obtained when the probe is substituted by a lateral magnetic dipole. Finally, we experimentally show that the reciprocal beam path configurations are equivalent and the probe can be used as a detector for the tangential magnetic field of light.

3.2 Results and Discussion

3.2.1 Hollow-Pyramid Aperture Probe as a tangential H_y Dipole Source: Intuitive Physical Justification

Often in near-field measurements it is convenient to approximate the probe to a sub-wavelength object for which the optical properties are well known in order to facilitate the interpretation of the images and the simulations of the experiments. For example, electric point dipole is used to model dielectric and apertureless metallic scattering SNOM probes, which are used for imaging of the electric field of light [9–12]. A split-ring probe has a strong magnetic dipole moment, normal to the sample surface [13].

Here, we propose that the SiO_2 hollow-pyramid metal coated aperture SNOM probes, which we have used in the previous chapter to image the lateral magnetic field of light, can be approximated by an H_y dipole—a tangential (lateral) y-polarized magnetic point dipole, where the y-direction is perpendicular to the light polarization direction and to the light propagation direction (Fig. 3.1).

The specific near-field probe investigated here consists of a hollow SiO_2 pyramid, coated with Al, with a sub-wavelength (100 nm) aperture at its apex (Fig. 3.1a). The structure of the probe was modeled following the information provided by the manufacturer (materials used, layer thicknesses, cone angle, and aperture size—Fig. 3.1b) [14].

The most straightforward simplification of the probe is to approximate the aperture in the probe to an aperture in a flat, perfectly conducting metal screen with

3.2 Results and Discussion

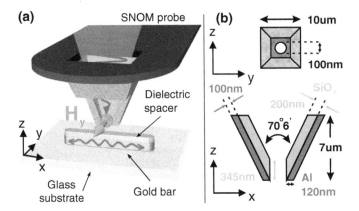

Fig. 3.1 a The hollow-pyramid probe of a SNOM can be approximated by an H_y magnetic dipole source (*green arrow*). The light is polarized in *x* direction. **b** Structure and dimensions of the pyramid probe, as provided by the probe supplier, are implemented in the FDTD simulations. The figure is not to scale

infinitesimal thickness. The latter is known as a Bethe-Bouwkamp aperture [15, 16]. It has been demonstrated that the light transmitted through such an aperture is similar to the combination of normal electric dipole and tangential magnetic dipole emissions [7, 17]. The direction of the magnetic dipole is determined by the polarization of the incident light.

The transmission through the hole is determined by the boundary conditions which the incident light has to satisfy at the air-metal interface. They require the tangential electric field E_{tan} and the normal magnetic field H_{norm} to be continuous at the interface between two materials. Inside a perfect conductor both the magnetic and the electric fields have to be zero, because a perfect metal has infinite conductivity, zero relaxation time and zero skin depth. Thus, the fulfillment of the continuity boundary condition requires that E_{tan} and H_{norm} components are zero also just outside the metal (Fig. 3.2a).

A perfect metal has zero relaxation time, which means that it reacts instantaneously to the incident electromagnetic wave. This means that we can assume illumination with any frequency as quasi-static. The quasi-static Maxwell equations allow decoupling the electric and the magnetic fields and treating separately the response of the metal to these fields. The free charges in the metal are redistributed by the oscillating electric field of the incident light, so that the E_{tan} component is canceled out and the electric field has only E_{norm} components close to the metal surface. Respectively, the oscillating magnetic field of the incident light is inducing screening currents inside the metal (red arrows, Fig. 3.2a). These currents effectively screen out the H_{norm} component and the magnetic field is dominantly parallel close to the metal surface. This intuitive perception is illustrated in Fig. 3.2a, where we sketch the electric and magnetic field distributions in the vicinity of the planar aperture. Rigorous calculations of those fields have been presented elsewhere [7, 17].

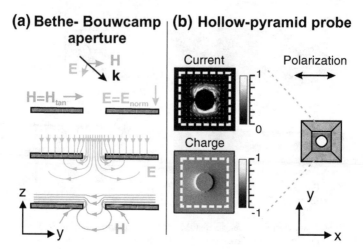

Fig. 3.2 A circular aperture probe can be simplified to a Bethe-Bouwkamp aperture—a hole in a perfectly conducting metal sheet. **a** Satisfying the boundary conditions results in the magnetic and electric field line distributions, hand-sketched in the middle and bottom panels. **b** Simulations of the real probe currents (*top*) and charges (*bottom*). The screening currents induced by the magnetic field of the incoming light are depicted in *red*. The profiles are taken 10 nm above the aperture at $\lambda = 1000$ nm

So far we have explained why H_{tan} and E_{norm} are the dominant field components near the air-metal interface. These components are also the main components present in the vicinity of the hole, because the hole is small compared to the wavelength of the light. Respectively, these are also the main components being transmitted through the hole. This is illustrated in Fig. 3.2a. Therefore, it can be assumed that the radiation transmitted through the hole can be effectively understood as the radiation of an H_{tan} and E_{norm} dipoles.

In the setup we are using, the incoming light is practically perpendicular to the plane of the metal screen and the hole, thus, the E_{norm} component is negligible. If an E_{norm} component is present, it will effectively act as an emitting E_z dipole. The radiation from such a dipole is not emitted along the z-axis, where our detector is located in the used measurement configuration. Therefore, even if it is present, the E_{norm} component will hardly contribute to the signal collected by the detector. If the scattered light is collected under a polarization resolved scheme, potentially the E_{norm} and E_{tan} components can also be obtained [7]. However in the current transmission configuration the probe is expected to behave only as a tangential magnetic dipole.

Next step is to simulate the real probe and verify that the above assumptions hold for the concrete geometry and parameters we use in our experiments. For that purpose we first perform finite-difference time-domain (FDTD) simulations of the generated currents and charges in the probe under an incident x-polarized plane wave (Lumerical © [18]). The above discussed screening currents are indeed generated in the probe, as illustrated with the red arrows in Fig. 3.2b, top. Similarly to the case of a Bethe-Bouwkamp aperture [15], these currents are expected to induce an effective

3.2 Results and Discussion

H_y magnetic dipole. Additionally, a dipolar charge polarization is observed at the apex of the probe—Fig. 3.2b, bottom.

To actually check whether the real probe can indeed be approximated by an effective H_y magnetic dipole, we need to simulate and compare the fields of the probe with the fields of such a dipole, which is discussed in the next section.

3.2.2 Correspondence Between the Fields of the Hollow-Pyramid Probe and a tangential H_y Dipole: Simulations

In this section we show that the simulated fields of the hollow-pyramid probe are indeed similar to the fields of an H_y magnetic dipole—Fig. 3.3.

The different non-zero electric and magnetic field components for an H_y dipole, as obtained from FDTD simulations, are plotted in panel a, where the white dot corresponds to the position of the dipole source in the simulation. Respectively, the

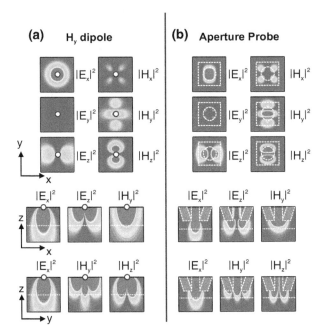

Fig. 3.3 a The close resemblance of the simulated near-field components of the magnetic dipole and **b** the probe indicates that the probe can be modeled as an H_y dipole source. The probe is outlined with the *white dashed lines* and the H_y dipole position is indicated with a *white circle*. The x-y field profiles are taken 30 nm below the probe and 130 nm below the dipoles (see horizontal *white dotted lines* in middle and bottom panels). The x-z and y-z profiles are taken through the middle of the probe/dipoles and only the non-zero field components are plotted. The simulations are performed at $\lambda = 1000$ nm

field components for the probe are shown in panel b, where the probe boundaries are indicated by the dashed white lines. The light incident on the probe is x-polarized with wavelength $\lambda = 1000$ nm. The profiles remain almost unchanged throughout a broad spectral range. The electric and the magnetic field plots have a common color scale. The x-y cross-sections are taken 30 nm below the aperture and 130 nm below the dipoles (see horizontal white dotted lines in the middle and bottom panels). The x-z and y-z cross-sections are taken through the aperture center.

The strong resemblance of the fields of the probe and the H_y dipole indicates that, in the near-field region, the aperture-probe can be approximated by such an H_y magnetic dipole. The simulated profiles in this section are in agreement with results obtained for gold-coated tapered fiber tips [19], a planar Bethe aperture [20] and a hole in a metal film at THz frequencies [21].

We have now confirmed by simulations our initial assumption that the aperture probe generates strongly resembling electromagnetic fields to those of an H_y dipole. Still, we need to verify that both the probe and the H_y dipole couple similarly to a sample.

3.2.3 Scanning of the Probe Over a Sample

In this section, we will show that the probe can be effectively substituted by the H_y dipole in a situation closely resembling the experiment, *i.e.* where the probe is being scanned over a metallic sample.

We have chosen to demonstrate this on a plasmonic antenna gold nanobar sample, in consistency with the structures studied and described in the previous chapters. The gold bar has a width of 70 nm and thickness of 50 nm. The gold is covered by a 30 nm thick SiO_2 capping layer [22]. This layer ensures that while scanning, the probe is not in direct conductive contact with the gold. The nanoantenna is supported by a SiO_2 substrate. As described before, when the metal bar is illuminated, different order surface plasmon resonance modes can be excited in it.

The surface plasmon resonances can be excited by disturbing the charge equilibrium in the metallic sample. However, the efficiency of the excitation depends on the type of excitation source, mainly its field distribution near the nanoantenna. In the previous section we have shown that the fields of the probe closely resemble those of an H_y dipole. Therefore, we can expect that both type of sources will excite the plasmon modes at the same locations along the bar.

To assess the excitation efficiency of the different plasmon modes we can compare the absorption of the metallic antenna when it is scanned by the probe and the H_y dipole—Fig. 3.4. Indeed, the absorption profiles obtained with the probe (b) are similar to the ones obtained with an H_y dipole (a). The different panels show the absorption profiles for the different order plasmon modes of the bar at the respective resonance wavelengths. The resonant wavelengths with and without the probe are different due to the proximity of the probe in the near-field of the sample [22]. The

3.2 Results and Discussion

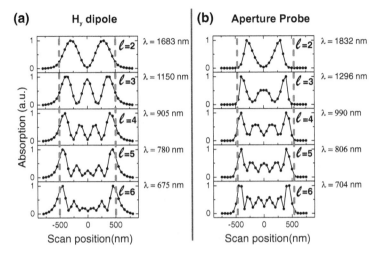

Fig. 3.4 b Similar absorption profiles are obtained when scanning a metallic bar with the probe and an with **a** an H_y magnetic dipole. This is illustrated for the different plasmon resonance modes in an $L = 1\,\mu\text{m}$ bar. The dipole is scanned at a height 130 nm above the bar and the probe is scanned in contact with the dielectric spacer on top of the bar. The edges of the bar are indicated with the *dashed red lines*. For clarity, each absorption profile is normalized independently

bar is scanned in 50 nm steps, indicated by the black dots along the x-axis. The y-axis gives the corresponding absorption at each of the scanning positions.

At the positions where a plasmon mode is most efficiently excited, a clear absorption peak is observed. We have normalized each of the modes for each of the sources separately in order to facilitate the comparison of the modes. Clearly, the position and shape of the absorption peaks is similar for the probe and the H_y dipole excitation. The finite dimensions of the probe are causing slight differences in the relative intensities and peak shapes. Additionally, other than H_y field components can be weakly transmitted through the probe, and contributing to the observed absorption cross-sections.

Based on these results, we can suggest that the plasmon antenna excitation by the probe can be qualitatively reproduced by substituting the probe with an H_y magnetic dipole. Such a substitution is practically very convenient, as it speeds up the simulation by about 30 times and reduces more than twice the volume of the simulation.

The validity of the substitution of the hollow-pyramid probe by a tangential magnetic dipole for scanning other types of samples, for example dielectric samples, remains to be realized. Such results have been reported in the literature for similar types of probes—metallized optical fiber probes [7, 8].

3.2.4 Experimental Evidence for Equivalence of Collection and Illumination Mode SNOM

So far we have discussed how the hollow-pyramid SNOM probe can be effectively approximated as an H_y dipole source. As discussed in the introduction, not only sources, but also detectors of the magnetic field of light are needed. According to the reciprocity theorem, in the far-field optical setups with inverted beam paths are equivalent [23]. However, there is a discussion about the validity of this theorem in the near-field and it has to be explicitly demonstrated for the concrete optical setup [24, 25]. In this section, we show experimentally that inverting the beam path in the hollow-pyramid SNOM setup results in similar experimental images, which suggests that the probe can be effectively used both as an optical H_y magnetic field source and detector.

In the illumination mode SNOM, the incoming light is focused by the objective at the top, through the tip, and the transmitted light is collected in the far field (Fig. 3.5a). The collection mode is reciprocal with inverted beam path—the sample is illuminated by a focused laser beam from the bottom at normal incidence through the substrate and the transmitted light is picked up through the probe and detected further by the objective at the top (Fig. 3.5b). This experiment has been demonstrated on a gold bar of length $L = 740$ nm, illuminated with a wavelength $\lambda = 785$ nm. The experimental transmission images show similar characteristic features in both

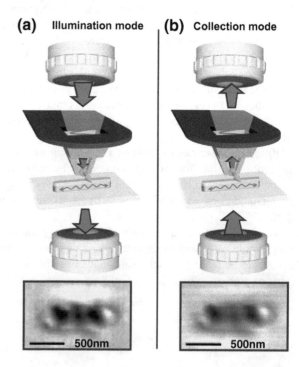

Fig. 3.5 Similar experimental near-field images are obtained in **a** illumination and **b** collection near-field imaging configurations. The sample is an $L = 740$ nm long bar, illuminated at $\lambda = 785$ nm, which results in the excitation of the $l = 3$ plasmon mode

measurement configurations—bottom panels in Fig. 3.5a, b. The regions of efficient plasmon mode excitation are observed as three dark spots, indicating the excitation of the third order antenna mode.

In the previous chapter we have demonstrated that the transmission image in illumination mode (Fig. 3.5a) can be associated with the H_y magnetic field distribution [22]. Using a similar setup with an optical fiber aperture probe, Kihm et al. [7] have shown that the experimental images in collection mode also visualize the H_y magnetic near-field for surface plasmon polaritons on a gold film. The images in collection mode have slightly reduced contrast, most probably due to slight misalignments as the alignment procedure is much more challenging from technical point of view.

Thus, the equivalence between illumination and collection modes is confirmed for this setup by the observed similarity of the experimental near-field images. The equivalence between the reciprocal configurations of an aperture optical fiber type of SNOM has also been discussed elsewhere [24, 25].

3.3 Conclusions

We suggest that, to first order approximation, a metal coated hollow-pyramid SNOM probe can be considered as a lateral magnetic dipole source and, reciprocally, lateral magnetic field detector. We support this suggestion by the simulated near-field profiles of the probe, which closely resemble the ones of a lateral magnetic dipole source. Additionally, the absorption profiles of a metallic sample scanned with the probe and with a lateral magnetic dipole show similar features. This allows us to confirm that the probe and the tangential magnetic point dipole source are coupling in a similar way to a metallic sample. To ultimately show that the probe can be substituted by a lateral magnetic dipole in a broader case, without the need to involve coupling to a metallic sample, additional studies are needed—namely, of a dielectric sample [7, 8].

The equivalence between the reciprocal use of the SNOM with reversed beam path is experimentally demonstrated, as similar images are obtained when using the probe as a lateral magnetic field source (illumination mode) and detector (collection mode). This has a practical importance as well, because the effects in the near-field are typically more intuitive to understand in the collection configuration when the probe is perceived as a magnetic field detector. However, typically the reciprocal illumination mode is more commonly used, as the setup is much easier to work with and align. The obtained results in this chapter open up new possibilities for performing much simpler and faster simulations and gain better understanding of the near-field images.

References

1. N. Liu, L.W. Fu, S. Kaiser, H. Schweizer, H. Giessen, Plasmonic building blocks for magnetic molecules in three-dimensional optical metamaterials. Adv. Mater. **20**, 3859–3865 (2008)
2. C. Rockstuhl, F. Lederer, C. Etrich, T. Pertsch, T. Scharf, Design of an artificial three-dimensional composite metamaterial with magnetic resonances in the visible range of the electromagnetic spectrum. Phys. Rev. Lett. **99**, 017401 (2007)
3. E. Xifré-Pérez, L. Shi, U. Tuzer, R. Fenollosa, F. Ramiro-Manzano, R. Quidant, F. Meseguer, Mirror-image-induced magnetic modes. ACS Nano **1**, 664–668 (2013)
4. D.R. Smith, W.J. Padilla, D.C. Vier, S.C. Nemat-Nasser, S. Schultz, Composite medium with simultaneously negative permeability and permittivity. Phys. Rev. Lett. **84**, 4184–4187 (2000)
5. C.M. Soukoulis, M. Wegener, Past achievements and future challenges in the development of three-dimensional photonic metamaterials. Nat. Photonics **5**, 523–530 (2011)
6. G. Dolling, C. Enkrich, M. Wegener, J.F. Zhou, C.M. Soukoulis, Cut-wire pairs and plate pairs as magnetic atoms for optical metamaterials. Opt. Lett. **30**, 3198–3200 (2005)
7. H.W. Kihm, J. Kim, S. Koo, J. Ahn, K. Ahn, K. Lee, N. Park, D.-S. Kim, Optical magnetic field mapping using a subwavelength aperture. Opt. Express **21**, 5625–5633 (2013)
8. B. le Feber, N. Rotenberg, D.M. Beggs, L. Kuipers, Simultaneous measurement of nanoscale electric and magnetic optical fields. Nat. Photonics **8**, 43–46 (2013)
9. L. Novotny, B. Hecht, *Principles of Nano-Optics* (Cambridge University Press, Cambridge, 2006)
10. J.-J. Greffet, R. Carminati, Image formation in near-field optics. Prog. Surf. Sci. **56**, 133 (1997)
11. J. Sun, P. Carney, J.C. Schotland, Strong tip effects in near-field scanning optical tomography. J. Appl. Phys. **102**, 103103 (2007)
12. A. García-Etxarri, I. Romero, J.F. García de Abajo, R. Hillenbrand, J. Aizpurua, Influence of the tip in near-field imaging of nanoparticle plasmonic modes: Weak and strong coupling regimes. Phys. Rev. B **79**, 125439 (2009)
13. M. Burresi, D. van Oosten, T. Kampfrath, H. Schoenmaker, R. Heideman, A. Leinse, L. Kuipers, Probing the magnetic field of light at optical frequencies. Science **326**, 550–553 (2009)
14. Witec Wissenschaftliche Instrumente und Technologie GmbH (2014). http://www.witec.de
15. H. Bethe, Theory of diffraction by small holes. Phys. Rev. **66**, 163–182 (1944)
16. C.J. Bouwkamp, Diffraction theory. Rep. Prog. Phys. **17**, 35 (1954)
17. A. Drezet, J.C. Woehl, S. Huant, Diffraction by a small aperture in conical geometry: application to metal-coated tips used in near-field scanning optical microscopy. Phys. Rev. E **65**, 046611 (2002)
18. Lumerical Solutions (2014). http://www.lumerical.com
19. P.-K. Wei, H.-L. Chou, Y.-R. Cheng, Y.-D. Yao, Near-field magneto-optical microscopy using surface-plasmon waves and the transverse magneto-optical Kerr effect. J. Appl. Phys. **98**, 093904 (2005)
20. A. Drezet, M.J. Nasse, S. Huant, J.C. Woehl, The optical near-field of an aperture tip. EPL (Europhys. Lett.) **66**, 41 (2004)
21. L. Guestin, A.J.L. Adam, J.R. Knab, M. Nagel, P.C.M. Planken, Influence of the dielectric substrateon the terahertz electric near-field of a hole in a metal. Opt. Express **17**, 17412–17425 (2009)
22. D. Denkova, N. Verellen, A.V. Silhanek, V.K. Valev, P. Van Dorpe, V.V. Moshchalkov, Mapping magnetic near-field distributions of plasmonic nanoantennas. ACS Nano **7**, 3168–3176 (2013)
23. M. Born, E. Wolf, *Principles of Optics*, 7th edn. (Cambridge U. Press, 1999)
24. K. Imura, H. Okamoto, Reciprocity in scanning near-field optical microscopy: Illumination and collection modes of transmission measurements. Opt. Lett. **31**, 1474–1476 (2006)
25. E. Méndez, J.-J. Greffet, R. Carminati, On the equivalence between the illumination and collection modes of the scanning near-field optical microscope. Opt. Commun. **142**, 7–13 (1997)

Chapter 4
Magnetic Near-Field Imaging of Increasingly Complex Plasmonic Antennas

Abstract In the previous two chapters we have verified that the aperture SNOM technique can be used for visualizing the lateral magnetic near-fields of metallic nanostructures. We have demonstrated the technique on geometrically simple structures—metallic bars as a proof of principle. Then we have discussed the underlying physics of this imaging concept. In this chapter we are expanding the method and applying it for imaging the magnetic near-fields of structures with geometries beyond simple bars. We first study other simple plasmonic nanoresonators, such as disks and rings and confirm that the technique is indeed imaging the lateral magnetic field for these geometries too. Then, we focus on more complex antennas, constructed from building blocks of different horizontal and vertical bars. For the studied structures, the magnetic near-field distributions of the complex resonators have been found to be a superposition of the magnetic near-fields of the individual constituting elements. These experimental results, explained and validated by numerical simulations, open new possibilities for engineering and characterization of complex plasmonic antennas with increased functionality.

4.1 Introduction

Currently, there is an increasing interest towards the field of plasmonics, due to the fascinating and somewhat unexpected optical properties, arising from the interaction of light with nanoscale metallic objects. The excitation of plasmons—collective oscillations of the free electrons in the metal, induced by the incident light, allows deep subwavelength concentration and localization of light in the near-field of the metal [1]. These enhanced electromagnetic fields have been proven to be technologically useful in hyper-sensitive bio and chemical detection [2–6], radiation control of

---

The results presented in this chapter are based on and reproduced with permission from:
D. Denkova, N. Verellen, A.V. Silhanek, P. Van Dorpe, V.V. Moshchalkov
Lateral magnetic near-field imaging of plasmonic nanoantennas with increasing complexity
Small, **10**, 1959–1966 (2014). Copyright © 2014, Wiley-VCH Verlag GmbH & Co. KGaA.

quantum emitters [7, 8], enhancement of various optical effects, such as non-linear [9, 10] and magneto-optic effects [11].

The main current research interest lies in both plasmonic structures with simple geometries (building blocks) [12–16] and complex structures with sophisticated interactions between their constituting elements [17–20]. The antennas with simple geometries are still of interest, because even the highly symmetric geometries (bar, disk, ring) already show non-trivial properties. In addition, they exhibit a simplified resonance mode structure and can be readily investigated using analytical methods.

Clearly, the complex antennas allow much more flexibility for artificially engineering their properties, but at the expense of more complicated interactions involved in the proper tuning of the antennas' functionalities. Recently, it has been demonstrated that complex antennas, based on simple building blocks, such as nanorods, have interesting applications, for example, in the nonlinear regime [21, 22], for directional light scattering [19, 23], optical switches [24], and for fine resonance tuning [25–27].

Although most of these studies provide detailed far-field characterization of the antennas and/or simulations of their electromagnetic near-fields, direct measurements of the near-field distributions are most often lacking. This is especially so for the magnetic field components, due to a weak magnetic interaction between light and matter at optical frequencies [28]. Acquiring knowledge about the magnetic field of light has become essential in the flourishing field of artificially engineered optical materials, so called metamaterials, where the enhanced response to the optical magnetic field can lead to extraordinary optical properties [29–32]. Thus, although there exist well-established methods for imaging the electric field components [14, 33–35], only recently experimental techniques for mapping the vertical [36], and later on, the lateral [37–39] magnetic field components, have been introduced. Therefore, detailed characterization of the different magnetic near-field components of plasmonic structures, the main building blocks for metamaterials [32], is an ongoing challenge.

Here, we present experimental lateral magnetic near-field distributions of various plasmonic antennas visualized by hollow-pyramid probe SNOM [37]. We have studied both simple antennas (bar, disk and ring) and more complex antennas, consisting of assembled horizontal and vertical bars in different geometrical configurations. Firstly, we confirm that the recently developed approach [37] for imaging the lateral magnetic field distribution in plasmon nanobars by hollow-pyramid probe SNOM, described in Chap. 2 and further developed in Chap. 3, is readily applicable also to more complex geometries. Secondly, we observe that for the studied structures, the magnetic near-field distributions of the complex antennas can be presented as a superposition of the magnetic near-field distributions of the individual non-interacting building blocks [27]. We confirm these results by finite difference time domain (FDTD) simulations [40].

4.2 Results and Discussion

The structures we have chosen to study are gold plasmonic antennas with various shapes, fabricated on a glass substrate (see Methods Sect. 4.4). The measurements are performed with a hollow-pyramid SNOM (WITec, alpha 300S) in transmission illumination configuration [41]. A detailed description of the setup is given in the Sect. 1.3.3. Shortly, a polarized laser beam is focused on the inner side of a hollow Al coated SiO_2 pyramid (Fig. 4.1), which has a subwavelength hole at its apex with a diameter of nominally 90 nm.

The horizontally (E_x) polarized incident light induces a dipolar charge separation at the apex of the probe, which in turn generates image charges in the nanoantenna underneath (Fig. 4.1, inset). This coupling results in the effective formation of a y-oriented magnetic dipole, which efficiently excites surface plasmons only at the lateral magnetic field ($|H_y|^2$) maxima of the antenna modes [37]. At those positions, the absorption in the plasmonic structures is enhanced and therefore a dip in the transmitted light intensity is observed. Consequently, the modulated transmitted light intensity, recorded while raster scanning the sample, provides a map of the lateral magnetic near-field distribution of the nanostructures.

The configuration we use necessitates a transparent substrate. In principle, if the excitation can be realized in a different way and the probe is used as a detector,

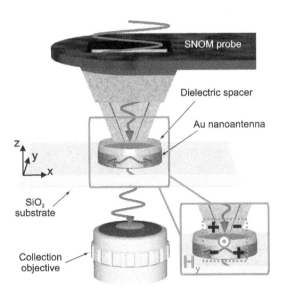

Fig. 4.1 Mapping of the magnetic $|H_y|^2$ near-field distribution of a plasmonic nanoantenna by a hollow-pyramid probe SNOM. The x-polarized light delivered through the subwavelength probe's aperture can locally excite surface plasmon resonances in the antenna. The interaction between the probe and the sample leads to the effective formation of an H_y dipole (*inset*), which efficiently excites the plasmon mode only at its $|H_y|^2$ maxima. As a result, the recorded transmitted light intensity during a sample raster scan gives the $|H_y|^2$ near-field map (Image not to scale.)

one could also measure the magnetic field on non-transparent substrates [39]. The substrate material will not significantly alter the plasmon modes of the sample, but will result in small spectral shifts. Since the raster scans are performed in contact scanning mode, to prevent a strong conductive coupling between the SNOM probe and the sample, which would otherwise substantially disturb the antenna properties, the metal particles are covered by a 30 nm thick dielectric resist layer. It behaves optically as SiO_2 and will red-shift the antenna resonance by a few tens of nanometers. Alternatively, the sample can be sputter-coated by a thin SiO_2 layer. The need for a dielectric spacer layer could be eliminated when using probes with a suitable spring constant for non-contact or tapping mode operation, which are currently not yet available [41]. More details on the measurement procedure, data interpretation and comparison with other near-field techniques are presented elsewhere [37].

Since the probe couples to the nanoantenna through the formation of an effective magnetic dipole, modelling of the full experimental system can be simplified by replacing the probe by a y-oriented magnetic point dipole source. Furthermore, although in the experiment the source (i.e. the SNOM probe) is scanned pixel by pixel to obtain the magnetic field map, the simulations allow to record the field inside the full simulation space. This allows us to obtain the $|H_y|^2$ near-field distribution map by simulating only one, or a few, dipole source positions, further reducing the simulation complexity and computational time. More details about the simulations are provided in the Methods Sect. 4.4 and in the Supporting Information Sect. 4.5.

4.2.1 Simple Antennas

First, we will discuss the results obtained for some of the simplest plasmonic resonator geometries, like a disk and a ring. Similar to previous reports on gold bars [37], we will show that also for the disk and the ring antenna geometries the experimentally obtained hollow-pyramid SNOM images reveal the $|H_y|^2$ field distribution (Fig. 4.2). Scanning electron microscope (SEM) images of the studied disk and ring structures are shown in the top row of Fig. 4.2. The disk has a diameter of 320 nm and the ring has an outer diameter of 470 nm and an inner diameter of 300 nm. Illuminating the disk and ring antenna results in collective oscillations of the free electrons in the metal, driven by the electromagnetic field of the incident light. Depending on the geometry and size of the particles, surface plasmon resonance modes can be excited at specific wavelengths. We will focus on the dipole plasmon mode. In the simulations the antennas are illuminated with the y-polarized magnetic dipole source discussed above, positioned 75 nm below the gold structure. The directions of the magnetic field and the corresponding electric field at the antenna are shown in Fig. 4.2b. The resulting dipole resonances occur at wavelengths of 1160 and 1980 nm for the disk and the ring, respectively. The corresponding charge distributions and magnetic and electric field components are plotted in Fig. 4.2. All electric and all magnetic field components are plotted in the same color scale, except for the weaker $|H_x|^2$, which is scaled up by a factor of 10. To ensure the best excitation efficiency

4.2 Results and Discussion

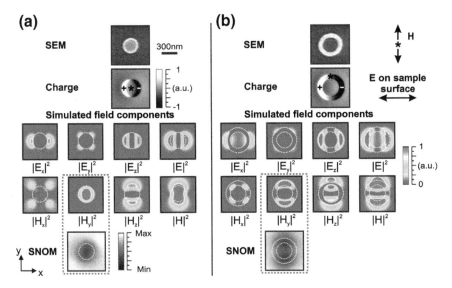

Fig. 4.2 a The SNOM images reveal the $|H_y|^2$ near-field distribution of the disk and **b** ring plasmonic structures. This is illustrated for the dipole mode, which is experimentally observed at $\lambda = 950$ nm for the disk and $\lambda = 1610$ nm for the ring. The simulated charge distribution and the corresponding field components are plotted at the wavelengths at which the dipole resonance is observed in the simulations ($\lambda = 1160$ nm for the disk and $\lambda = 1980$ nm for the ring). The antennas are excited via an H_y magnetic dipole source, positioned 75 nm below the antenna (*black star*). The directions of the source's magnetic field and the corresponding electric field at the antenna are shown in (**b**). Dimensions of the structures are: disk diameter 320 nm; ring inner diameter 300 nm, ring outer diameter 470 nm. The x-y cross sections are taken through the middle of the structures for the charge plots and 60 nm above the top surface of the structure for the field components plots

in the simulations, the source was positioned at the lateral magnetic field maxima. The positions of the exciting dipoles are indicated with black stars in Fig. 4.2.

The experimental SNOM images, both for the disk and the ring, are shown in the bottom row of Fig. 4.2. The dipole mode of the disk shows a slightly elliptical dark spot, the region corresponding to strong plasmon excitation by the effective magnetic dipole of the probe. For the ring, on the other hand, two separated dark lobes are observed. From a comparison with the simulated field components, it can be seen that these SNOM maps correspond to the $|H_y|^2$ magnetic field distributions. The dipole resonance wavelengths at which the experimental images are obtained are $\lambda = 950$ nm and $\lambda = 1610$ nm for the disk and the ring respectively. The blue-shifts in the experimental resonance wavelengths compared to the simulated ones are mainly caused by the probe-sample interaction, which is not accounted for in the simulation [37].

4.2.2 Complex Antennas Consisting of Assembled Bars

In this section, we have chosen to build-up complex antennas from already well characterized individual resonators—gold bars [37]. We study the interactions between the individual bar elements and the consequent effect on the magnetic near-field distributions. More specifically, we will first consider a structure consisting of two parallel bars with varying gap between them. Afterwards, we will discuss more elaborated structures, composed of both horizontal and vertical bars with varying size. As mentioned in the introduction, similar structures have recently attracted a lot of interest because of the promising applications for optical switches [24], enhancement of non-linear effects [21, 22] and fine resonance tuning [25–27].

Two Parallel Bars

The building block of the parallel bars structure is a bar with length $L = 1020$ nm and width $W = 220$ nm. The simulated absorption spectrum of such a bar is shown in Fig. 4.3a, top panel. The different peaks in the spectrum correspond to the different order standing wave-like plasmon cavity modes of the bar, indicated with the colored numbers above each peak. We will mainly focus on the $l = 3$ mode of this bar, which occurs at $\lambda = 1135$ nm (dashed line in the spectra in Fig. 4.3a). The corresponding simulated lateral magnetic field distribution for this mode showing three magnetic field maxima (Fig. 4.3b, top panel) is clearly reproduced in the experimental image where the corresponding three dark spots are observed (Fig. 4.3c, top panel, plotted at the experimental $l = 3$ resonance wavelength $\lambda = 785$ nm). This is in agreements with previous results [37], therefore, for clarity, other field components of the nanobar are not shown. As for the disk and the ring simulations, here again the bar is excited by an H_y point dipole, positioned at an $|H_y|^2$ magnetic field maximum and indicated with a black star.

We can now add an identical bar, parallel to the initial one characterized above. First, we position it at a distance at which both bars are barely interacting, separated by a gap, G, of 380 nm. The weak interaction between the two bars is detectable in this case as the small spectral changes in Fig. 4.3a, second row from top. Even though the excitation is still realized by an H_y dipole positioned at the left $|H_y|^2$ field maximum of the top bar, the $l = 3$ mode of the bottom bar is already slightly excited, as it can be seen in the corresponding $|H_y|^2$ field profile in panel b. The plasmon oscillations in the individual bars are occurring out of phase (Fig. 4.8).

In the simulations the two bars are excited by placing the dipole source at one single position (black star), while the experimental maps result from a complete raster scan of the source across the visualized area and recording of the transmitted light intensity at each source position. For this reason, in the simulated images one of the bars has a weaker contrast, while in the experimentally obtained images the two bars have identical contrast. The simulated $|H_y|^2$ profiles and the experimental images are always plotted at the resonance wavelengths of the single, non-interacting bar to facilitate tracking of the changes in the basic bar structure field profiles. The contrast in the experimental images (Fig. 4.3c) has been individually optimized for

4.2 Results and Discussion

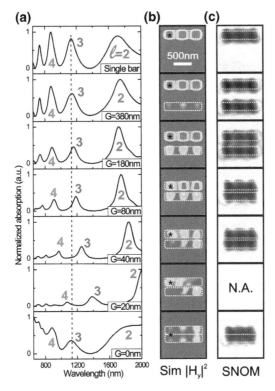

Fig. 4.3 The $|H_y|^2$ near-field distribution of a horizontal bar at its resonance wavelength (plotted here for the $l = 3$ mode, three $|H_y|^2$ maxima) is not significantly disturbed by the presence of an additional horizontal bar, for a broad variation of interparticle gaps. However, for gaps between $G = 0$ nm and $G = 40$ nm, at the same wavelength, the higher order mode is excited ($l = 4$ mode, four $|H_y|^2$ maxima), due to the stronger interaction between the two antennas. **a** Normalized simulated absorption spectra, where the mode order is indicated with *colored numbers*. **b** Corresponding simulated $|H_y|^2$ field distributions for the $l = 3$ mode ($\lambda = 1135$ nm, *dashed line* in panel (**a**)), excited via an H_y dipole positioned at the *black star*. The spectra in (**a**) are obtained for the same source position. **c** Experimental SNOM maps for x-polarized incident light confirming the simulations presented in (**b**), plotted at the experimental $l = 3$ resonance wavelength ($\lambda = 785$ nm). Dimensions of the bars are: length $L = 1020$ nm, width $W = 220$ nm. x-y cross-sections of the field components are taken 60 nm above the top surface of the structure

each image. The simulated $|H_y|^2$ maps (Fig. 4.3b) are plotted in the same color scale. The spectra in Fig. 4.3a are normalized to their maximum, after subtracting the base line.

Bringing the two parallel bars closer together leads to more significant changes in the absorption spectrum. Decreasing the gap distance (from top to bottom in Fig. 4.3) leads to stronger red-shift of the plasmon resonance wavelengths. However, since the modes of the bars are clearly defined, relatively broad and almost not overlapping, the $l = 3$ mode field distribution of the building blocks (three magnetic field maxima) in the combined parallel bar structure is preserved for $\lambda = 1135$ nm wavelength down to a gap distance of $G = 40$ nm. These observations are also confirmed by the experimental maps in Fig. 4.3c, where the corresponding three dark spots are observed. Additionally, the strength of the mode excitation is still significant for all studied gaps, as can be seen from comparison of the $|H_y|^2$ field distribution images in Fig. 4.3b. A representative image of the other field components for a gap of $G = 180$ nm is shown in the Supporting Information, Sect. 4.5.

For interparticle gap smaller than $G = 40$ nm already a more significant spectral shift is observed and consequently the mode excited at $\lambda = 1135$ nm is the higher order $l = 4$ mode exhibiting four magnetic field maxima (Fig. 4.3b, $G = 20$ nm, second panel from bottom to top). Unfortunately, the fabrication of a structure for SNOM measurements with such a gap separation has not been successful, therefore, no experimental image is provided for this case. Touching of the two parallel bars ($G = 0$ nm) leads to the effective formation of a bar, identical to the initial basic building block, but twice wider. Therefore, the exciting dipole for this structure was positioned in the middle of the composed bar. Since for horizontal illumination, the strongest resonance excited in the structure is defined by the horizontal dimensions of the cavity, the wider bar and the original one should yield similar responses. This is illustrated by the last panel of Fig. 4.3, where the $l = 3$ mode is restored at $\lambda = 1135$ nm for the simulations and $\lambda = 785$ nm for the measurements.

Thus, for the studied parallel bars structures, there is a narrow region of gap separations (below $G = 40$ nm) for which the spectrum is significantly red-shifted due to the mutual interaction between the two bars. In these structures, at the resonant wavelength of the $l = 3$ mode of the non-interacting bar, the next order $l = 4$ mode is excited in the interacting parallel bars. For all other separations between the two bars, including touching bars, the interaction strength between the building blocks is not sufficient to significantly shift the spectrum and the $l = 3$ mode is preserved in the parallel bars.

We would like to emphasize that these results should not be straightforwardly generalized for other geometries, especially in quantitative aspect, since for building blocks more complex than the bars, various factors, such as mode overlapping, sharper resonances with high quality factors, Fano resonances etc., may cause much more sensitive shifting of the resonance positions with the gap variation [17, 42].

Assembled Perpendicular and Vertical Bars

The next step in building more complex resonator geometries is to introduce vertical bars as building blocks. The antennas studied in this section are different combinations of merged horizontal and vertical bars (Fig. 4.4). For these antennas, similarly to the two parallel bars structure, we will demonstrate that the hollow-pyramid SNOM images correspond to the $|H_y|^2$ near-field distribution. Furthermore, we will show that the field distribution of the complex antennas looks like a direct superposition of the fields of the non-interacting building blocks.

The building blocks of the complex antennas are vertical and horizontal short and long bars (Fig. 4.4). The long bars have a length of $L = 1020$ nm and all bars have a width of $W = 220$ nm. The length of the short bars is chosen so, as to place their $l = 2$

4.2 Results and Discussion

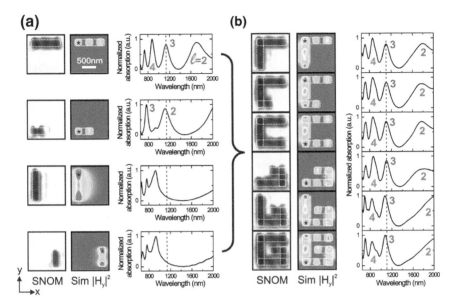

Fig. 4.4 The near-fields of the studied complex antennas are a superposition of the near-fields of their non-interacting building blocks. **a** Experimental SNOM images, simulated $|H_y|^2$ field distribution and corresponding absorption spectra for the building block antennas. **b** Same for the complex antennas composed by properly assembling the building blocks in (**a**). The width of all bars is $W = 220$ nm and the length of the long bars is $L = 1020$ nm. The short bar has $L = 620$ nm in the simulations and $L = 520$ nm in the measurements. The simulated images are obtained at the resonance wavelength of the $l = 3$ mode of the long horizontal bars ($\lambda = 1135$ nm), indicated by a *dashed line* in the spectra. The procedure for obtaining the simulated images is explained in the main text and in the Supporting Information, Sect. 4.5. The *x*-*y* cross-sections of the field components are taken 60 nm above the top surface of the structure. The spectra are obtained for the dipole positioned at the *black star*. Resonance plasmon mode orders are indicated with *colored numbers*. The experimental SNOM images are obtained with *x*-polarized light at $\lambda = 785$ nm, which is the experimental $l = 3$ resonance wavelength for the long bar and $l = 2$ for the short one

plasmon resonance at the same wavelength as the $l = 3$ resonance of the long bars ($L = 620$ nm at $\lambda = 1135$ nm for the simulations and $L = 520$ nm at $\lambda = 785$ nm for the measurements). The simulated $|H_y|^2$ magnetic field distributions, absorption spectra, and the experimental SNOM images, for the building blocks and for the complex antennas constructed from them, are shown in Fig. 4.4a, b, respectively.

The simulated field maps for the horizontal bars have been obtained in the same way as for the other studied structures so far, i.e. by positioning a dipole near the edge of the bar, at the maximum of the antenna's magnetic field distribution. However, for the vertical bars and the complex antennas, the full near-field map of the whole structures cannot be visualized by positioning a local source at just one position (see Supporting Information, Sect. 4.5). Therefore, to obtain a full map of the near-field distribution around the whole structure, several simulations with different positions of the source have to be performed and the resulting near-field images added up. A detailed description of how these images are composed, including source positions, is given in the Supporting Information, Sect. 4.5. The agreement between the experimental images and the simulated $|H_y|^2$ field distribution indicates that the hollow-pyramid SNOM is indeed mapping the $|H_y|^2$ lateral magnetic field distributions also for antennas with more complex geometries.

Similarly to the study for the two parallel bars structures, also here the resonance in a certain assembled building block is not significantly perturbed by the presence of other building blocks around it. The spectra of the complex antennas, excited at the source locations indicated with the black stars are given in Fig. 4.4b. The resonance mode orders are indicated near the corresponding spectral peaks with colored numbers. Clearly, the spectral features (peak position and relative intensities) of the horizontal long bar building blocks are preserved in all studied configurations and are barely disturbed by the presence of other building blocks around them.

Accordingly, the field distributions of the complex antennas shown in this Figure are a superposition of the fields of the non-interacting building blocks in Fig. 4.4a. For example, the field distribution of the long horizontal bar building block in Fig. 4.4a consists of three magnetic field maxima and the field distribution of the long vertical one shows a relatively homogeneous magnetic field distribution. Combining those two structures in one antenna (Fig. 4.4b, first row) results in a new antenna, in which both building blocks have preserved their individual, non-interacting field distributions. This is observed both in the experimental images and the simulations. A similar effect has also been recently reported by Jeyaram et al. [27].

4.3 Conclusions

In this chapter we present experimental maps of the lateral magnetic near-field distribution of plasmonic antennas with different geometries, including both simple resonators and more complex ones, consisting of different bars used as building blocks. We demonstrate that mapping of the magnetic near-field distribution with an aperture SNOM technique is not restricted to nanorod antennas [37], but is also applicable to other antenna geometries.

Notably, the magnetic near-field of the studied complex antennas turns out to be a superposition of the magnetic fields of the non-interacting building blocks. With the help of numerical simulations, we propose an explanation of this effect—for the studied structures, even though a weak interaction between the building blocks is observable, the simplicity of the resonance mode structure and the clear spectral separation of the different modes in the bars result in insignificant perturbations of the near field profile at the resonance wavelength by the presence of the additional building blocks.

This opens up interesting opportunities, such as close packing on all-optical chips, where the proximity of other antennas will not significantly alter the pre-engineered response of the individual self-standing antennas. These findings should facilitate studying of very complex antennas, which can now be described as a simple

4.3 Conclusions

superposition of their elementary building blocks' near-field distributions. Additionally, combining the here presented lateral magnetic field maps with vertical magnetic field maps, which can be obtained by other techniques [36], paves the way for full 3D characterization of nanoscale magnetic light-matter interactions.

4.4 Methods

4.4.1 Sample Preparation

The nanoantennas are supported by a glass substrate coated with 10 nm of indium tin oxide (ITO) and a thin Ti adhesion layer. The nanostructures consist of sputtered gold and are patterned using electron beam lithography on a negative tone hydrogen silsesquioxane (HSQ) resist with subsequent ion milling. A 30 nm thick dielectric resist layer is left on top of the structures. This layer behaves optically as SiO_2 and prevents a strong conductive coupling between the SNOM probe and the sample in contact-scanning mode. This allows us to do fast contact mode scanning of the sample without drastically altering the plasmonic properties.

4.4.2 Simulations

Simulations are performed with a commercial finite-difference time-domain (FDTD) solver Lumerical © [40]. The glass substrate and the resist layer on top of the gold antennas are included in the simulations (refractive index n = 1.4). The simulations were performed with perfectly matched layer (PML) boundaries. The structures were excited by one or several magnetic point dipole sources. The permittivity of Au was taken from Ref. [43]. The ITO and Ti layers are not included in the simulations. More detailed information about the performed simulations is provided in the next Sect. 4.5.

4.5 Supporting Information

The supporting information describes and justifies the procedure followed to construct the simulated H_y near-field distribution maps in Fig. 4.4 of this chapter.

It has been demonstrated previously that in the SNOM measurements, an H_y magnetic dipole between the hollow-pyramid SNOM probe and the metal structure is effectively formed [37]. Therefore, in the simulations we have locally excited the plasmonic antennas by such a dipole source, thus simulating the effect of the aperture SNOM probe. We have used the commercially available finite-difference time-domain (FDTD) Lumerical software to perform the simulations [40]. The magnetic field direction is shown in Fig. 4.5a. As depicted there, the charges in the metal antennas, illuminated by the dipole source, feel a horizontally polarized electric field (x-direction).

In the SNOM experiments, the source is raster scanned pixel by pixel and the transmitted light intensity is recorded at each position to form the final SNOM image. We have shown that for our experimental configuration the transmitted light intensity map is equivalent to the $|H_y|^2$ near-field distribution map [37]. While the transmitted light intensity provides information only about a single point of the image, in the simulations we can record the $|H_y|^2$ field in the whole simulation space. This allows us to obtain a larger region of the $|H_y|^2$ near-field distribution map with one simulation, considerably reducing the simulation complexity and computational time.

For certain configurations the $|H_y|^2$ near-field map of the whole plasmonic resonator can be obtained by locally exciting the antennas at a single position (Sect. 4.5.1). However, to be able to map the full near-field distribution in more general cases, we introduce an alternative procedure (Sects. 4.5.2 and 4.5.3).

For the sake of completeness, the different field components of a representative complex antenna shown in Fig. 4.4 are presented in Sect. 4.5.4.

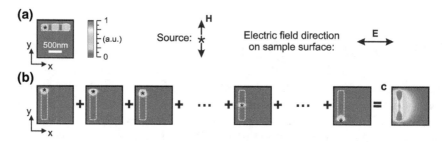

Fig. 4.5 **a** For a horizontal bar, illuminated by an H_y dipole source (*black star*), the charges are driven by a horizontally polarized electric field and the whole bar is playing the role of a resonator. Thus, the $|H_y|^2$ near-field distribution of the whole bar can be imaged by positioning the source at a single location (*black star*). **b** For a vertical bar, whose charges are excited by a horizontally polarized electric field, only the charges in the vicinity of the source oscillate along the width of the bar and respectively only the field close to the source can be visualized. Therefore, to obtain the full near-field map, the source is scanned in 50 nm steps along the bar length. **c** The near-field map of the whole bar is obtained by summing the fields at the different positions of the source, shown in (**b**). The *x-y* cross sections are plotted 60 nm above the gold structures

4.5 Supporting Information

4.5.1 Simulation of the $|H_y|^2$ Near-Field Map of an Elementary Horizontal Bar

Let us consider a bar, illuminated by the H_y local dipole source, described above. The charges in the vicinity of the source start to oscillate synchronously, following the direction of the electric field (with a certain phase delay). If the bar is horizontal and the direction of the electric field is horizontal too, practically the whole bar plays the role of a resonator for the oscillating charges. Thus, at the resonance frequencies of the bar, all charges oscillate together to form a standing wave resonating in the whole bar. Therefore, even though the source is local, the full near-field map of the plasmon mode in the bar can be visualized by positioning the source at one single location (Fig. 4.5a).

4.5.2 Simulation of the $|H_y|^2$ Near-Field Map of an Elementary Vertical Bar

Let us now discuss a vertical bar, locally illuminated by the H_y dipole source. The charges in the vicinity of the source are oscillating following the horizontal electric field, but in this case the resonator is formed along the width of the bar. This is illustrated in Fig. 4.5b, where the field distributions of a vertical bar for a few positions of the source (black star) are shown. Clearly, only the charges in the vicinity of the source are strongly oscillating, while the charges distant from the source do not participate in the oscillation. Consequently, only the field distribution in the vicinity of the source can be visualized. Therefore, under these circumstances, to perform a simulation reproducing the full map of the near-field distribution along the whole bar we need to scan the source and obtain information from different source positions, analogous to the images captured in the experiments.

In order to reduce the simulation time and complexity, instead a raster scan of the whole image (as in the measurements), a scan of the dipole source in 50 nm steps along the bar length is performed. To obtain the full $|H_y|^2$ map (Fig. 4.5c) we sum up all the images taken at the different positions of the source. This procedure has been used to obtain the $|H_y|^2$ field distribution of the vertical bars in Fig. 4.4.

4.5.3 Simulation of the $|H_y|^2$ Near-Field Map of Complex Antennas

Similarly to the vertical bars, in the more complex structures in Fig. 4.4b, the plasmon resonance modes of the whole structure cannot be simultaneously excited by positioning the dipole source at one single location. Hence, a procedure similar to the one explained above for the vertical bars needs to be applied. It is illustrated in

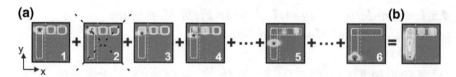

Fig. 4.6 a The whole complex antenna (Fig. 4.4b, first row) cannot be excited by positioning the dipole source (shown in Fig. 4.5a at a single location). Therefore, simulations performed at several source positions (*black stars*) are needed to obtain the magnetic near-field distribution of the whole structure. **b** The full $|H_y|^2$ near-field map of the antenna is obtained by summing the fields at the different positions of the source, shown in (**a**). For the final image, to avoid artificial increase of the signal from the *horizontal bar*, only one position of the source at the *horizontal bar* is taken into account and only the fields to the left of the *grey dashed line* are summed. The x-y cross sections are plotted 60 nm above the gold structures

detail in Fig. 4.6, where the $|H_y|^2$ near-field profiles for the different locations of the source (black star, Fig. 4.6a) are summed up to generate the full magnetic field distribution (Fig. 4.6b).

The source positions in Fig. 4.6a are chosen consistently with the procedures described in Sects. 4.5.1 and 4.5.2. In panel 1, Fig. 4.6a the source is located at the intersection between the horizontal and the vertical bar. As explained in Sects. 4.5.1 and 4.5.2, the whole horizontal bar is excited, while the vertical bar away from the source is not. Thus, here again, to obtain the field along the vertical bar, we move the source down along the middle of the vertical bar in 50 nm steps. However, the second position, at 50 nm down from the center of the horizontal bar is not taken into account (Fig. 4.6a, panel 2). Since there, the source is still located within the boundaries of the horizontal bar, it would excite again the whole horizontal bar resulting in artificial increase of the signal from the horizontal bar in the final image.

The next simulation added to the final image is the one where the source is located at 100 nm down from the center of the horizontal bar (Fig. 4.6a, panel 3). However, since at this position the horizontal bar is still excited, to avoid again artificial enhancement of the horizontal bar's field in the final image, only the fields to the left of the grey dashed line were added up.

This procedure is repeated in 50 nm steps until the end of the bar. For all those positions, again, only the fields to the left of the dashed grey line are added to form the final image (Fig. 4.6b). Images with a source position at 400 and 900 nm distance from the middle of the horizontal bar are shown in panels 4 and 5 in Fig. 4.6a. All images in Fig. 4.6a are with the same color scale. The color scale in Fig. 4.6b has been scaled down by a factor of three. The field maps of the other complex antennas in Fig. 4.4b are constructed following the same procedure. The dipole source positions which were taken into account for the formation of these images are shown in Fig. 4.7.

This simulation method with partially scanning the source over the nanoantenna allows us to closely reproduce the SNOM experiment while maintaining reasonable simulation time and computing power. Note that while the distribution of magnetic

4.5 Supporting Information 77

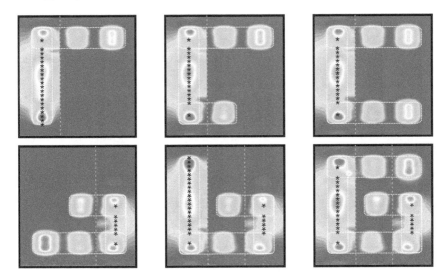

Fig. 4.7 The whole complex antennas (Fig. 4.4b) cannot be excited by positioning the dipole source at a single location. Thus, the full $|H_y|^2$ near-field maps of the antennas are obtained by summing the fields at different positions of the source, indicated by the *black stars*. To avoid artificial increase of the signal from the horizontal bar, only one position of the source at the horizontal bar is taken into account and only the fields to the left of the *grey dashed line* are summed up for the left vertical bars and accordingly, to the right of the respective *dashed grey lines* for the right vertical bars. The x-y cross sections are plotted 60 nm above the gold structures

field maxima and minima is accurately reproduced, their relative intensities inherently depend on the exact sum series of source positions and are, therefore, not exact.

4.5.4 Electromagnetic Field Components of a Representative Complex Antenna

The different field components of a representative complex antenna are shown in Fig. 4.8. The oscillations of the plasmons in the two rods are occurring out of phase.

Fig. 4.8 Different field components of a representative complex antenna shown in Fig. 4.3, gap $G = 180$ nm. The x-y cross sections are plotted 60 nm above the gold structures

References

1. J.A. Schuller, E.S. Barnard, W. Cai, Y.C. Jun, J.S. White, M.L. Brongersma, Plasmonics for extreme light concentration and manipulation. Nat. Mater. **9**, 193–204 (2010)
2. A.G. Brolo, Plasmonics for future biosensors. Nat. Photonics **6**, 709–713 (2012)
3. K. Lee, J. Irudayaraj, Correct spectral conversion between surface-enhanced Raman and plasmon resonance scattering from nanoparticle dimers for single-molecule detection. Small **9**, 1106–1115 (2013)
4. A.V. Kabashin, P. Evans, S. Pastkovsky, W. Hendren, G.A. Wurtz, R. Atkinson, R. Pollard, V.A. Podolskiy, A.V. Zayats, Plasmonic nanorod metamaterials for biosensing. Nat. Mater. **8**, 867–871 (2009)
5. J.N. Anker, W.P. Hall, O. Lyandres, N.C. Shah, J. Zhao, R.P. Van Duyne, Biosensing with plasmonic nanosensors. Nat. Mater. **7**, 442–453 (2008)
6. J. Homola, Surface plasmon resonance sensors for detection of chemical and biological species. Chem. Rev. **108**, 462–493 (2008)
7. A.G. Curto, T.H. Taminiau, G. Volpe, M.P. Kreuzer, R. Quidant, N.F. van Hulst, Multipolar radiation of quantum emitters with nanowire optical antennas. Nat. Commun. **4**, 1750 (2013)
8. M.K. Schmidt, S. Mackowski, J. Aizpurua, Control of single emitter radiation by polarization- and position-dependent activation of dark antenna modes. Opt. Lett. **37**, 1017–1019 (2012)
9. V.K. Valev, D. Denkova, X. Zheng, A.I. Kuznetsov, C. Reinhardt, B.N. Chichkov, G. Tsutsumanova, E.J. Osley, V. Petkov, B. De Clercq, A.V. Silhanek, Y. Jeyaram, V. Volskiy, P.A. Warburton, G.A.E. Vandenbosch, S. Russev, O.A. Aktsipetrov, M. Ameloot, V.V. Moshchalkov, T. Verbiest, Plasmon-enhanced sub-wavelength laser ablation: plasmonic nanojets. Adv. Mater. **24**, OP29–OP35 (2012)
10. M. Kauranen, A.V. Zayats, Nonlinear plasmonics. Nat. Photonics **6**, 737–748 (2012)
11. J.Y. Chin, T. Steinle, T. Wehlus, D. Dregely, T. Weiss, V.I. Belotelov, B. Stritzker, H. Giessen, Nonreciprocal plasmonics enables giant enhancement of thin-film faraday rotation. Nat. Commun. **4**, 1599 (2013)
12. G.W. Bryant, F.J. García de Abajo, J. Aizpurua, Mapping the plasmon resonances of metallic nanoantennas. Nano Lett. **8**, 631–636 (2008)
13. R. Esteban, R. Vogelgesang, J. Dorfmüller, A. Dmitriev, C. Rockstuhl, C. Etrich, K. Kern, Direct near-field optical imaging of higher order plasmonic resonances. Nano Lett. **8**, 3155–3159 (2008)
14. J. Dorfmüller, R. Vogelgesang, T.R. Weitz, C. Rockstuhl, C. Etrich, T. Pertsch, F. Lederer, K. Kern, Fabry-Pérot resonances in one-dimensional plasmonic nanostructures. Nano Lett. **9**, 2372–2377 (2009)
15. J. Dorfmüller, R. Vogelgesang, W. Khunsin, C. Rockstuhl, C. Etrich, K. Kern, Plasmonic nanowire antennas: experiment, simulation, and theory. Nano Lett. **10**, 3596–3603 (2010)
16. E.R. Encina, E.A. Coronado, Resonance conditions for multipole plasmon excitations in noble metal nanorods. J. Phys. Chem. C **111**, 16796–16801 (2007)
17. Z.-J. Yang, Z.-S. Zhang, L.-H. Zhang, Q.-Q. Li, Z.-H. Hao, Q.-Q. Wang, Fano resonances in dipole-quadrupole plasmon coupling nanorod dimers. Opt. Lett. **36**, 1542–1544 (2011)
18. Z.-G. Dong, H. Liu, M.-X. Xu, T. Li, S.-M. Wang, S.-N. Zhu, X. Zhang, Plasmonically induced transparent magnetic resonance in a metallic metamaterial composed of asymmetric double bars. Opt. Express **18**, 18229–18234 (2010)
19. J. Dorfmüller, D. Dregely, M. Esslinger, W. Khunsin, R. Vogelgesang, K. Kern, H. Giessen, Near-field dynamics of optical Yagi-Uda nanoantennas. Nano Lett. **11**, 2819–2824 (2011)
20. T. Kosako, Y. Kadoya, H.F. Hofmann, Directional control of light by a nano-optical Yagi-Uda antenna. Nat. Photonics **4**, 312–315 (2010)
21. V.K. Valev, B. De Clercq, C.G. Biris, X. Zheng, S. Vandendriessche, M. Hojeij, D. Denkova, Y. Jeyaram, N.C. Panoiu, Y. Ekinci, A.V. Silhanek, V. Volskiy, G.A.E. Vandenbosch, M. Ameloot, V.V. Moshchalkov, T. Verbiest, Distributing the optical near-field tor efficient field-enhancements in nanostructures. Adv. Mater. **24**, OP208–OP215 (2012)

References

22. G.F. Walsh, L. Dal Negro, Enhanced second harmonic generation by photonic-plasmonic fano-type coupling in nanoplasmonic arrays. Nano Lett. **13**, 3111–3117 (2013)
23. D. Vercruysse, Y. Sonnefraud, N. Verellen, F.B. Fuchs, G. Di Martino, L. Lagae, V.V. Moshchalkov, S.A. Maier, P. Van Dorpe, Unidirectional side scattering of light by a single-element nanoantenna. Nano Lett. **13**, 3843–3849 (2013)
24. V.K. Valev, A.V. Silhanek, B. De Clercq, W. Gillijns, Y. Jeyaram, X. Zheng, V. Volskiy, O.A. Aktsipetrov, G.A.E. Vandenbosch, M. Ameloot, V.V. Moshchalkov, T. Verbiest, U-shaped switches for optical information processing at the nanoscale. Small **7**, 2573–2576 (2011)
25. N. Verellen, P. Van Dorpe, C. Huang, K. Lodewijks, G.A.E. Vandenbosch, L. Lagae, V.V. Moshchalkov, Plasmon line shaping using nanocrosses for high sensitivity localized surface plasmon resonance sensing. Nano Lett. **11**, 391–397 (2011)
26. J.-S. Huang, J. Kern, P. Geisler, P. Weinmann, M. Kamp, A. Forchel, P. Biagioni, B. Hecht, Mode imaging and selection in strongly coupled nanoantennas. Nano Lett. **10**, 2106–2110 (2010)
27. Y. Jeyaram, N. Verellen, X. Zheng, A.V. Silhanek, M. Hojeij, B. Terhalle, Y. Ekinci, V.K. Valev, G. Vandenbosch, V.V. Moshchalkov, Rendering dark modes bright by using asymmetric split ring resonators. Opt. Express **21**, 15464–15474 (2013)
28. L.D. Landau, E.M. Lifshitz, *Electrodynamics of Continuous Media* (Pergamon, Oxford, 1960)
29. V.M. Shalaev, Optical negative-index metamaterials. Nat. Photonics **1**, 41–48 (2007)
30. T. Grosjean, M. Mivelle, F.I. Baida, G.W. Burr, U.C. Fischer, Diabolo nanoantenna for enhancing and confining the magnetic optical field. Nano Lett. **11**, 1009–1013 (2011)
31. S. Koo, M.S. Kumar, J. Shin, D. Kim, N. Park, Extraordinary magnetic field enhancement with metallic nanowire: role of surface impedance in Babinet's principle for sub-skin-depth regime. Phys. Rev. Lett. **103**, 263901 (2009)
32. A.V. Kildishev, A. Boltasseva, V.M. Shalaev, Planar photonics with metasurfaces. Science **339**, 1289 (2013)
33. S.I. Bozhevolnyi, Near-field mapping of surface polariton fields. J. Microsc. **202**, 313–319 (2001)
34. J.-S. Bouillard, S. Vilain, W. Dickson, A.V. Zayats, Hyperspectral imaging with scanning near-field optical microscopy: applications in plasmonics. Opt. Express **18**, 16513 (2010)
35. P. Alonso-Gonzalez, M. Schnell, P. Sarriugarte, H. Sobhani, C. Wu, N. Arju, A. Khanikaev, F. Golmar, P. Albella, L. Arzubiaga, F. Casanova, L.E. Hueso, P. Nordlander, G. Shvets, R. Hillenbrand, Real-space mapping of Fano interference in plasmonic metamolecules. Nano Lett. **11**, 3922–3926 (2011)
36. M. Burresi, D. van Oosten, T. Kampfrath, H. Schoenmaker, R. Heideman, A. Leinse, L. Kuipers, Probing the magnetic field of light at optical frequencies. Science **326**, 550–553 (2009)
37. D. Denkova, N. Verellen, A.V. Silhanek, V.K. Valev, P. Van Dorpe, V.V. Moshchalkov, Mapping magnetic near-field distributions of plasmonic nanoantennas. ACS Nano **7**, 3168–3176 (2013)
38. H.W. Kihm, J. Kim, S. Koo, J. Ahn, K. Ahn, K. Lee, N. Park, D.-S. Kim, Optical magnetic field mapping using a subwavelength aperture. Opt. Express **21**, 5625–5633 (2013)
39. B. le Feber, N. Rotenberg, D.M. Beggs, L. Kuipers, Simultaneous measurement of nanoscale electric and magnetic optical fields. Nat. Photonics **8**, 43–46 (2013)
40. Lumerical Solutions (2014). http://www.lumerical.com
41. Witec Wissenschaftliche Instrumente und Technologie GmbH (2014). http://www.witec.de
42. M.K. Gupta, T. König, R. Near, D. Nepal, L.F. Drummy, S. Biswas, S. Naik, R.A. Vaia, M.A. El-Sayed, V.V. Tsukruk, Surface assembly and plasmonic properties in strongly coupled segmented gold nanorods. Small **9**, 2979–2990 (2013)
43. P.B. Johnson, R.W. Christy, Optical constants of the noble metals. Phys. Rev. B **6**, 4370–4379 (1972)

Chapter 5
Conclusions and Outlook

The field of optics is dramatically expanding its scope during the last years. Photonic nanomaterials, and in particular plasmonic nanoantennas, enable light and matter manipulation at the nanoscale. The ability of plasmonic structures to confine light to nanometer dimensions has made them promising building blocks for ultrahigh sensitivity bio- and chemical-sensors, all-optical computing devices and photodetectors, to name a few.

The functionalities of those plasmonic devices are mainly determined by the light-matter interactions in the nanovolumes of, and around, metal nanostructures. Thus, optimizing and broadening of those functionalities require detailed characterization of the specific parameters of that interaction, such as charge, current, electric and magnetic field distributions of the plasmonic modes.

Since light is an electromagnetic wave, it interacts with materials via its oscillating electric and magnetic fields. However, the photonics community has long disregarded the magnetic component of the electromagnetic field of light, focusing on the exploration of the electric component. The main reason is that at optical frequencies natural materials do not exhibit strong magnetic response, so the magnetic light-matter interactions are, firstly, not of essential importance for those materials, and secondly—very difficult to excite and measure. However, new classes of artificial nanomaterials, such as plasmonic nanoantennas and metamaterials are often designed to have a very strong magnetic response. Therefore, in the recent years, a need emerged for both optical magnetic sources and detectors for the magnetic field of light and significant efforts have been invested in this direction.

5.1 Conclusions

The work presented in this thesis contributes to the quest for magnetic sensors/emitters at optical frequencies.

We suggest that the aperture of a hollow-pyramid probe scanning near-field optical microscope (SNOM) can be considered as a local optical magnetic dipole source and, reciprocally, detector for the magnetic field of light. This is indicated by simulating and comparing the near-field profiles of the probe with those of electric and magnetic dipole sources. In addition to the probe itself, we study the interaction between the probe and a gold nanobar sample. We show, both experimentally and theoretically, that the probe is indeed imaging the tangential magnetic field distribution of the different plasmonic modes excited in gold bars with different length.

Thus, the proposed hollow-pyramid aperture probe as an optical magnetic field source and detector contributes to the task of complete characterization of nanoscale devices, by providing an opportunity to explore magnetic light-matter interactions. Additionally, the possibility to substitute the probe by a magnetic dipole source in the simulations greatly facilitates the calculations by significantly reducing the simulations' complexity, time and memory requirements.

The reciprocity theorem in optics states that optical configurations with reversed light paths, meaning interchanged source and detector, are equivalent. This theorem is undoubtedly proven to be valid in the far-field, however it's applicability to the near-field has long been discussed. Although there are some reports confirming it's applicability for near-field measurements, the fine experimental details are of crucial importance. Our experiments indicate that the theorem should be valid also in the near-field zone for the specific SNOM probe and configuration which we use. From practical point of view the equivalence between the configurations with reciprocal beam paths is very convenient. From one side, the near-field interactions and images are typically more intuitive to understand when the probe is perceived as a magnetic field detector. On the other side however, from technical point of view, the setup is much easier to work with and align in the reciprocal mode, where the probe is used for excitation and the detection is realized in the far-field.

After demonstrating the applicability of the aperture hollow-pyramid SNOM to image the lateral magnetic field of light on a basic plasmonic nanobar sample, we apply the technique to study plasmonic antennas with different geometries. We visualize the magnetic near-field distributions of different antenna topologies including complex antennas, consisting of different elementary building blocks. Similar nanoantennas have recently attracted a lot of interest for sensor applications and for enhancement of non-linear and magneto-optical effects. However, direct experimental visualization of their magnetic near-fields was, until now, lacking. Therefore, the obtained magnetic near-field maps are essential for understanding and further developing reinforced magnetic light-matter interactions in plasmonic nanomaterials.

The near-field profiles reveal that the magnetic field distributions of the studied complex antennas are a superposition of the magnetic near-fields of the individual, non-interacting constituting elements. With the help of numerical simulations, we propose an explanation of this effect—even though a weak interaction between the building blocks is observable, the simplicity of the resonance mode structure and the clear spectral separation of the different modes in the bars result in insignificant perturbations of the near field profile at the resonance wavelength by the presence of the additional building blocks. This opens up interesting new opportunities, both

5.1 Conclusions

technologically and scientifically, for example, for closer packing on all-optical chips, optical chemical- and bio-sensors, and for simplifying complex antennas studies.

In the time frame in which our work was conducted, alternative methods for imaging the tangential magnetic field of light have been suggested by other groups. Namely, the use of other circular aperture SNOM probes based on optical fibers, has been proposed. Compared to the optical fiber probes, the hollow-pyramid probes which we use are more robust and provide higher-resolution topography images simultaneously with the optical magnetic field images. Additionally, the probes allow operation in a broad wavelength range and negligible polarization effect on the transmitted light. The drawback is that for metallic samples, additional precautions (for example a dielectric spacer on top of the sample) have to be foreseen to prevent direct conductive contact between the probe and the sample.

5.2 Outlook

Modern science and engineering have reached levels which allow us to design new optical materials with unprecedented properties, impossible to find in nature—so called metamaterials. The creation of "magic"-like objects, for example an invisibility cloak, is reaching out of the science-fiction books and nearing closer and closer to reality. One of the main factors opening the door for the fascinating properties of metamaterials is that, unlike natural materials, metamaterials are capable to strongly interact with the magnetic field of the light. For the technological progress and future development of this field, it is of crucial importance to characterize the various aspects of the light-matter interaction with nano-scale resolution. One of the last-standing challenges in that respect is, precisely, to gain experimental access to the magnetic component of the light at optical frequencies.

In this thesis, we have addressed the quest for magnetic field detectors/sources at optical frequencies by demonstrating how the hollow-pyramid aperture probe of a scanning near-field optical microscope can be used as a lateral magnetic dipole source and, respectively, as a detector for nano-imaging the tangential magnetic field component of the light. As suggested for similar imaging techniques, other components of the electromagnetic field should be accessible by upgrading the microscope with additional detectors and polarizers. Polarization-resolved images, obtained by a detector collecting the scattered light should allow access to the tangential and vertical electric field components.

The aperture probes, combined with the other developed probes for imaging the different components of the electromagnetic field of light, make possible the measurement of the complete nanoscale electromagnetic field vector. This is a powerful tool to drive advances in the field of metamaterials, and to unravel the interplay between the geometrical structures and the resulting optical behaviour.

In this thesis we have focused our efforts on studying plasmonic nanoantenna structures. The demonstrated imaging of the lateral magnetic field of light should be readily applicable also to other samples, for example dielectric samples such as photonic crystal cavities. The technique would be especially beneficial for metamaterials or other structures, specially developed for enhancing the light-matter interactions. Additionally, the aperture near-field probe can be used to explore various fundamental processes in nano-objects, such as, for example, a molecule undergoing a magnetic dipole transition.

Curriculum Vitae

Denitza Denkova
Born December 24th, 1985
Sofia, Bulgaria

2010–2014: Ph.D. research at the Department of Physics and Astronomy, Institute for Nanoscale Physics and Chemistry (INPAC), KU Leuven, Leuven, Belgium. Supervisor: Prof. Dr. V.V. Moshchalkov, co-supervisors: Dr. N. Verellen and Dr. V.K. Valev.

2008–2010: Master student at Sofia University, Sofia, Bulgaria. Junior researcher at the Ellipsometry group.
Thesis subject: "Development and characterization of Cathodoluminescence setup and preliminary studies on ZnO nanoparticles and surface plasmons in metal structures". Supervisor: Dr. Stoyan Russev.

2004–2008: Bachelor student at Sofia University, Sofia, Bulgaria. Junior researcher at the Ellipsometry group and the Solid State Spectroscopy group.
Thesis subject: "Gate oxide breakdown—True root cause determination by breakdown features".
Prepared as a joint project between: Sofia University, Bulgaria; Melexis Ltd., Bulgaria; Melexis Ltd., Germany; X-FAB Germany; Fraunhofer Institute for Mechanics of Materials, Germany

1999–2004: Student at the National High School of Mathematics and Natural Sciences, Sofia, Bulgaria. Major Physics.

Articles in International Peer-Reviewed Academic Journals

1. **D. Denkova**, N. Verellen, A.V. Silhanek, V.K. Valev, P. Van Dorpe, V.V. Moshchalkov
   Mapping magnetic near-field distributions of plasmonic nanoantennas
   ACS Nano **7**, 3168 (2013).
2. **D. Denkova**, N. Verellen, A.V. Silhanek, P. Van Dorpe, V.V. Moshchalkov
   Lateral magnetic near-field imaging of plasmonic nanoantennas with increasing complexity
   Small, **10**, 1959–1966 (2014).
3. **D. Denkova**, N. Verellen, A.V. Silhanek, P. Van Dorpe, V.V. Moshchalkov
   Near-field aperture-probe as a magnetic dipole source and optical magnetic field detector
   Submitted to arXiv:1406.7827 [Physics. Optics] (2014).
4. V.K. Valev, **D. Denkova**, X. Zheng, A.I. Kuznetsov, C. Reinhardt, B.N. Chichkov, G. Tsutsumanova, E.J. Osley, V. Petkov, B. De Clercq, A.V. Silhanek, Y. Jeyaram, V. Volskiy, P.A. Warburton, G.A.E. Vandenbosch, S. Russev, O.A. Aktsipetrov, M. Ameloot, V.V. Moshchalkov, T. Verbiest
   Plasmon-enhanced sub-wavelength laser ablation: plasmonic nanojets
   Adv. Mater. **24**, OP29–OP35 (2012).
5. N. Verellen, F. López-Tejeira, R. Paniagua-Dominguez, D. Vercruysse, **D. Denkova**, L. Lagae, P. Van Dorpe, V.V. Moshchalkov, J. Sanchez-Gil
   Mode parity-controlled Fano- and Lorentz-like line shapes arising in plasmonic nanorods
   Nano Lett., **14**, 2322–2329 (2014).
6. V.K. Valev, B. De Clercq, C.G. Biris, X. Zheng, S. Vandendriessche, M. Hojeij, **D. Denkova**, Y. Jeyaram, N.C. Panoiu, Y. Ekinci, A.V. Silhanek, V. Volskiy, G.A.E. Vandenbosch, M. Ameloot, V.V. Moshchalkov, T. Verbiest
   Distributing the optical near field for efficient field-enhancements in nanostructures
   Adv. Mater. **24**, OP208–OP215 (2012).
7. V.K. Valev, B. De Clercq, X. Zheng, **D. Denkova**, E.J. Osley, S. Vandendriessche, A.V. Silhanek, V. Volskiy, P.A. Warburton, G.A.E. Vandenbosch, M. Ameloot, V.V. Moshchalkov, T. Verbiest
   The role of chiral local field enhancements below the resolution limit of Second Harmonic Generation microscopy
   Opt. Express **20**, 256–264 (2012).
8. V.K. Valev, A.V. Silhanek, Y. Jeyaram, **D. Denkova**, B. De Clercq, V. Petkov, X. Zheng, V. Volskiy, W. Gillijns, G.A.E. Vandenbosch, O.A. Aktsipetrov, M. Ameloot, V.V. Moshchalkov and T. Verbiest
   Hotspot Decorations map plasmonic patterns with the resolution of scanning probe techniques
   Phys. Rev. Lett. **106**, 226803 (2011).

9. N. Verellen, **D. Denkova**, V.K. Valev, B. De Clercq, A.V. Silhanek, J.J. Baumberg, M. Ameloot, P. Van Dorpe, V.V. Moshchalkov
   Two-photon luminescence of gold nanorods mediated by higher order antennas modes of odd and even parity
   In preparation.
10. X. Zheng, N. Verellen, V.K. Valev, V. Volskiy, **D. Denkova**, L.O. Herrmann, C. Blejean, J.J. Baumberg, A.V. Silhanek, G.A.E. Vandenbosch, V.V. Moshchalkov
    The roles of plasmonic materials and geometry in shaping the optical response of nanontennas: an n-port network model
    In preparation.

Conference Contributions

1. **D. Denkova**, N. Verellen, A.V. Silhanek, V.K. Valev, P. Van Dorpe, V.V. Moshchalkov
   Mapping magnetic near-field distributions of plasmonic nanoantennas (Oral presentation)
   Frontiers in Optics 2013. Orlando (U.S.A.), 6–10 October 2013.
   Abstract FTu5D.1.
2. **D. Denkova**, N. Verellen, A.V. Silhanek, V.K. Valev, P. Van Dorpe, and V.V. Moshchalkov
   Optical magnetic near-field imaging in plasmonic nanoantennas (Poster presentation)
   International Photochemistry Conference (ICP) Conference. Leuven, Belgium, 21–26 July 2013.
3. **D. Denkova**, N. Verellen, A.V. Silhanek, V.K. Valev, P. Van Dorpe, and V.V. Moshchalkov
   Near-field imaging of surface plasmon resonances in gold nanobars with hollow-pyramid probes (Poster presentation)
   Near-Field Optics (NFO12) Conference. Donostia—San Sebastian, Basque Country, Spain, 3–7 September 2012.
4. **D. Denkova**, N. Verellen, A.V. Silhanek, V.K. Valev, P. Van Dorpe, and V.V. Moshchalkov
   Near-field imaging of surface plasmon resonances in gold nanobars with hollow-pyramid probes (Oral presentation)
   SPIE Photonics Europe. Brussels, Belgium, 16–19 April 2012.
5. N. Verellen, **D. Denkova**, A.V. Silhanek, V.K. Valev, P. Van Dorpe, V.V. Moshchalkov
   Near-field imaging of plasmonic nanoantennas: revealing magnetic field distributions (Oral presentation)
   Eighth International Conference on Vortex Matter in Nanostructured Superconductors. Rhodes, Greece, 21–26 September 2013.

6. N. Verellen, **D. Denkova**, Y. Jeyaram, M. Shestakov, V. Tikhomirov, V.K. Valev, A.V. Silhanek, P. Van Dorpe, V.V. Moshchalkov
   Light-matter interactions mediated by nanoscale confinement in plasmonic resonators. (Invited talk)
   Advanced Optoelectronics & Lasers Conference and International Workshop on Nonlinear Photonics, NLP 2013. Sudak, Ukraine, 9–13 September 2013.
   Published in Proceedings: 10.1109/NLP.2013.6646375, pp. 17–18.
7. N. Verellen, **D. Denkova**, A.V. Silhanek, V.K. Valev, P. Van Dorpe, V.V. Moshchalkov
   Mapping optical near-fields in gold nanobars with hollow-pyramid SNOM probes (Oral presentation)
   E-MRS (European Materials Research Society) 2012 spring meeting. Strasbourg, France, 14–18 May 2012.
8. V.V. Moshchalkov, N. Verellen, V. Tikhomirov, **D. Denkova**, Y. Jeyaram, M.V. Shestakov, V.K. Valev, A.V. Silhanek, P. Van Dorpe
   Light-matter interactions in plasmonic nanoresonators (Invited plenary talk)
   Symposium on plasmon-based chemistry and physics (pre-conference of International Photochemistry Conference (ICP 2013)). Leuven, Belgium, 19–20 July 2013.
9. N. Verellen, F. López-Tejeira, R. Paniagua-Domínguez, D. Vercruysse, **D. Denkova**, L. Lagae, P. Van Dorpe, V.V. Moshchalkov, and J.A. Sánchez-Gil
   Fano- and Lorentz-like resonances in plasmonic nanorods (Oral presentation)
   META 14 Conference. Singapore, 20–23 May 2014.
10. N. Verellen, F. López-Tejeira, R. Paniagua-Domínguez, D. Vercruysse, **D. Denkova**, L. Lagae, P. Van Dorpe, V.V. Moshchalkov, and J.A. Sánchez-Gil
    Mode parity-controlled Fano and Lorentz resonances in plasmonic nanorods (Accepted for poster presentation, paper number: 9163–86)
    SPIE NanoScience + Engineering, Optics and Photonics Conference. San Diego, USA, 17–21 August 2014.
11. N. Verellen, **D. Denkova**, A.V. Silhanek, V.K. Valev, P. Van Dorpe, V.V. Moshchalkov
    Aperture-SNOM reveals plasmonic magnetic near-fields (Accepted for oral presentation, paper number: 9169–17)
    SPIE NanoScience + Engineering, Optics and Photonics Conference. San Diego, USA, 17–21 August 2014.

CPSIA information can be obtained
at www.ICGtesting.com
Printed in the USA
LVHW02*1212300918
591916LV00004B/641/P